T0297921

RANDOM PROCESSES
BY EXAMPLE

RANDOM PROCESSES
BY EXAMPLE

Mikhail Lifshits

St. Petersburg State University, Russia
& Linköping University, Sweden

 **World Scientific**

NEW JERSEY · LONDON · SINGAPORE · BEIJING · SHANGHAI · HONG KONG · TAIPEI · CHENNAI

Published by

World Scientific Publishing Co. Pte. Ltd.

5 Toh Tuck Link, Singapore 596224

USA office: 27 Warren Street, Suite 401-402, Hackensack, NJ 07601

UK office: 57 Shelton Street, Covent Garden, London WC2H 9HE

British Library Cataloguing-in-Publication Data
A catalogue record for this book is available from the British Library.

RANDOM PROCESSES BY EXAMPLE

ISBN 978-981-4522-28-1

Printed in Singapore

Preface

The story of these lectures is somewhat unusual. Few years ago I was reading a research manuscript by I. Kaj and M.S. Taqqu [49] that handled the limit behavior of "teletraffic systems". Clearly, this was a deep and important piece of modern mathematics brilliantly representing a wave of interest to the subject, see e.g. [8, 10], [13, 14, 48], [69, 87], [70, 71, 108], to mention just a few.

What apparently passed less appreciated, was a tremendous value of teletraffic models for teaching and learning. In a simple and intuitively trivial model, a minor tuning of few parameters leads to different workload regimes – Wiener process, fractional Brownian motion, stable Lévy process, as well as to some less commonly known ones, called by Kaj and Taqqu "Telecom processes".

Simplicity of the dependence mechanism used in the model enables to get a clear understanding both of long and short range dependence phenomena. On the other hand, the model shows how light or heavy distribution tails lead either to continuous process, or to processes with jumps in the limiting regime. Furthermore, the existing multivariate extensions of the model may have completely different applied meaning, e.g. in medicine.

The "only" problem for implementation of the teletraffic model into teaching is that the vocabulary it is based on is somewhat peripheral for common courses of random processes. I mean here, among others, independently scattered measures and related integrals, stable random variables and processes, and, to a certain extent, the variety of Gaussian processes.

Therefore, before sharing with my students the treasures of knowledge encoded in teletraffic models, I had to explain them the preliminaries, providing something like "additional chapters of random process theory", – although anyone familiar with applied probability models will confirm that it is not an "additional stuff" but a necessary toolbox you have to be armed with.

What you see below is a result of author's modest efforts to mark up the way to theoretical and applied stochastic models that he loves so much.

Therefore, the goal of these lectures is twofold.

First, we aim to expose the mathematical tools necessary for understanding and working with a broad class of applied stochastic models, while presenting the material in a quick and condensed way.

The toolbox includes Gaussian processes, independently scattered measures (with Gaussian white noise, Poisson random measures, and stable random measures as main examples), stochastic integrals, compound Poisson, infinitely divisible and stable distributions.

Second, we illustrate many general concepts by handling the "infinite source teletraffic model" in a setting due to I. Kaj and M. S. Taqqu. This is what I. Norros [77] called "teletraffic as a stochastic playground" – but taken rather seriously. Hopefully, handling in depth of full scale examples (unusual for a textbook) brings into routine learning a touch of research work.

The author did his best in helping the reader

- to become familiar with a wide class of key random processes;
- to understand how probability theory works in an important applied problem;
- to get an idea of variety of limit theorems, especially for random processes.

The primary intended readership are PhD or Masters students, as well as researchers working in pure or applied mathematics.

In university teaching, one can build a one-semester advanced course upon this book. Such courses were recently given by the author at St.Petersburg State University (Russia).

M. A. Lifshits

Acknowledgments

First thanks go to my *alma mater*, Matmech Department of St. Petersburg State University, where I had a chance to test in teaching some parts of the material. I am also extremely grateful to Matematiska Institutionen (MAI) of Linköping University for providing me with excellent conditions for working on the manuscript.

The work was supported by RFBR grant 13-01-00172 and SPbSU grant 6.38.672.2013.

I am grateful to I. Norros for useful advise, to D. Egorov for pointing out numerous flaws in early manuscript versions and to E. Kosarevskaya for the help with preparation of pictures.

Special thanks are due to World Scientific representatives, Dr. Lim Swee Cheng and Kwong Lai Fun for their continuous help and encouragement so important at all stages of editorial process.

Contents

Chapter 1

Preliminaries

1 Random Variables: a Summary

This section only summarizes, in a crash course style, basic probabilistic notions needed in the sequel. We refer to Shiryaev [93] for the true complete exposition.

1.1 *Probability space, events, independence*

A *probability space* is a mathematical model for real-world situations or processes with the known set of *possible* outcomes. We do not know which outcome will occur in the real world but we define, according to our perception of the process, the *probabilities* of realization for certain groups of outcomes.

The probability space includes three components: a *sample space* Ω, a *sigma-field* of *events* $\mathcal{A}$, and a *probability* function $\mathbb{P}$. The sample space Ω is a set of all possible outcomes for the process we are modelling.

To certain subsets of Ω, called events, we assign a probability to get the corresponding outcome. In other words, for a subset $A \subset \Omega$ the probability $\mathbb{P}(A)$ is a number describing the chance to get an outcome from A. An event is considered to have "happened" when the outcome being a member of the event is observed.

Notice that, in general, probability is *not* assigned to *all* subsets of the sample space.

Let denote $\mathcal{A}$ the collection of all events (the subsets to which a probability is assigned). Then probability is a function $\mathbb{P} : \mathcal{A} \mapsto [0, 1]$.

The triplet $(\Omega, \mathcal{A}, \mathbb{P})$ is a *measure space* which means the following. First, $\mathcal{A}$ is a sigma-field, which simply means that a) the entire space Ω is an event; b) if A is an event, then the complement $\overline{A} := \Omega \backslash A$ is an event, and c) any finite or countable union of events is an event.

Second, $\mathbb{P}$ is a *probability measure*. The basic property of a measure – countable additivity – means that for any finite or countable set of *disjoint* events we have

$$\mathbb{P}\{\cup_j A_j\} = \sum_j \mathbb{P}(A_j).$$

Our probability is normalized so that $\mathbb{P}(\Omega) = 1$.

Events A and B are called *independent* if

$$\mathbb{P}(AB) = \mathbb{P}(A) \cdot \mathbb{P}(B).$$

Independence of three events A, B, C writes as

$$\mathbb{P}(AB) = \mathbb{P}(A) \cdot \mathbb{P}(B),$$
$$\mathbb{P}(AC) = \mathbb{P}(A) \cdot \mathbb{P}(C),$$
$$\mathbb{P}(BC) = \mathbb{P}(B) \cdot \mathbb{P}(C),$$
$$\mathbb{P}(ABC) = \mathbb{P}(A) \cdot \mathbb{P}(B) \cdot \mathbb{P}(C).$$

In general case, independence of several events means that the probability of intersection of any subset of these events equals to the product of the corresponding probabilities.

Exercise 1.1. For any event B let $\overline{B} := \Omega \backslash B$ denote the complementary event. Prove that:

- If A, B are independent, then $A, \overline{B}$ are independent.
- If A, B are independent, then $\overline{A}, \overline{B}$ are independent.
- If $\mathbb{P}(A) = 0$, then $\forall B$ events A, B are independent.
- If $\mathbb{P}(A) = 1$, then $\forall B$ events A, B are independent.

The following example recalls that independence of several events is not the same as their pairwise independence.

Example 1.2. (Bernstein pyramid). Take a four-faced regular pyramid. Assume that the faces of the pyramid are colored: one face is red, the second is white, the third is blue, while the fourth one bears a white-red-blue flag. We throw the pyramid on the table and look at its lower face after throwing. We assume, of course, that every face appears with equal probability $\frac{1}{4}$. Consider now three events:

$$R := \{\text{Red color appears on the chosen face}\};$$
$$W := \{\text{White color appears on the chosen face}\};$$
$$B := \{\text{Blue color appears on the chosen face}\}.$$

Clearly, $\mathbb{P}(R) = \mathbb{P}(W) = \mathbb{P}(B) = \frac{1}{2}$ while $\mathbb{P}(RW) = \mathbb{P}(RB) = \mathbb{P}(WB) = \frac{1}{4}$ and $\mathbb{P}(RWB) = \frac{1}{4}$. Obviously, these three events are pairwise independent but taken all together they are dependent.

1.2 *Random variables and their distributions*

In the following we denote $\mathbb{R}$ or $\mathbb{R}^1$ the set of real numbers, while $\mathbb{R}^n$ stands for Euclidean space of dimension n. We denote $\mathcal{B}^n$ Borel sigma-field, the smallest sigma-field of subsets of $\mathbb{R}^n$ containing all open sets. We also write $\mathcal{B} := \mathcal{B}^1$.

A *random variable* is any measurable function $X : (\Omega, \mathcal{A}) \mapsto (\mathbb{R}, \mathcal{B})$. Here "measurable" just means that for any $B \in \mathcal{B}$ we have

$$X^{-1}(B) := \{\omega \in \Omega : X(\omega) \in B\} \in \mathcal{A}.$$

The *distribution* of X is a measure on $(\mathbb{R}, \mathcal{B})$ defined by

$$P_X(B) := \mathbb{P}(X \in B), \qquad B \in \mathcal{B}.$$

The distribution of X is uniquely characterized by the *distribution function*

$$F_X(r) := P_X(-\infty, r] = \mathbb{P}(X \le r), \qquad r \in \mathbb{R}.$$

Random variables are called *identically distributed* or *equidistributed*, if their distributions coincide.

To a random variable X we associate the sigma-field

$$\mathcal{A}_X := \{X^{-1}(B), \ B \in \mathcal{B}\} \subset \mathcal{A}.$$

One may determine whether any event from $\mathcal{A}_X$ occurred, just by knowing the value of X.

Random variables $X_1, \ldots, X_n$ are called *independent*, if any events

$$A_1 \in \mathcal{A}_{X_1}, \ldots, A_n \in \mathcal{A}_{X_n}$$

are independent. In other words, for any sets $B_1, \ldots, B_n \in \mathcal{B}$ the events $\{X_1 \in B_1\}, \ldots, \{X_n \in B_n\}$ must be independent.

It is customary in Probability not to distinguish two random variables X, Y such that $\mathbb{P}(\omega : X(\omega) \neq Y(\omega)) = 0$. Moreover, by defining a random variable, it is admissible to leave it undefined on a set (event) of probability zero. Such uncertainty does not affect the distribution or any probabilistic property of the variable.

In applications one mostly works either with discrete or absolutely continuous distributions.

One calls the distribution of a random variable X, and the random variable X itself, *discrete* if X takes values from a finite or countable set $(x_j) \subset \mathbb{R}$. In this case the distribution is given by positive probabilities $p_j := \mathbb{P}(X = x_j)$ that must satisfy the equation

$$\sum_j p_j = 1.$$

For example, a random variable X has *Poisson distribution* $\mathcal{P}(a)$ with parameter $a > 0$ iff

$$\mathbb{P}(X = k) = e^{-a} \frac{a^k}{k!}, \qquad \forall \, k = 0, 1, 2, 3, \ldots$$

Exercise 1.3. Let X_1 and X_2 be independent variables with Poisson distributions $\mathcal{P}(a_1)$ and $\mathcal{P}(a_2)$ respectively. Then the distribution of their sum $X_1 + X_2$ also is Poisson distribution $\mathcal{P}(a_1 + a_2)$.

One calls the distribution of a random variable X *absolutely continuous*, if there exists a *distribution density* $p_X(\cdot)$, i.e. a non-negative integrable function on $\mathbb{R}$ such that

$$\mathbb{P}(X \in B) = P_X(B) = \int_B p_X(r) dr, \qquad B \in \mathcal{B}.$$

Clearly, we have

$$\int_{-\infty}^{\infty} p_X(r) dr = \mathbb{P}\{X \in \mathbb{R}\} = \mathbb{P}(\Omega) = 1.$$

For example, a random variable is said to have a *normal distribution* $P_X := \mathcal{N}(a, \sigma^2)$ with parameters $a \in \mathbb{R}, \sigma^2 > 0$, if it has a density

$$p_X(r) = \frac{1}{\sqrt{2\pi}\,\sigma} \, e^{-\frac{(r-a)^2}{2\sigma^2}}, \qquad r \in \mathbb{R}.$$

The family of normal distributions is completed by discrete single-point distributions $\mathcal{N}(a, 0)$, corresponding to constant random variables X such that $P_X(a) = \mathbb{P}(X = a) = 1$.

The distribution $\mathcal{N}(0, 1)$ is called *standard normal*. Any normal distribution can be reduced to or obtained from the standard normal distribution by a linear transformation. Namely, if $P_X = \mathcal{N}(a, \sigma^2)$, then $P_{\frac{X-a}{\sigma}} = \mathcal{N}(0, 1)$. Conversely, if $P_X = \mathcal{N}(0, 1)$, then

$$P_{\sigma X + a} = \mathcal{N}(a, \sigma^2). \tag{1.1}$$

It is clear now that for the family of normal distributions σ is a *scale parameter*, while a is a *shift parameter*.

The standard normal distribution is often given by the distribution function

$$F_X(r) := \mathbb{P}(X \le r) = \frac{1}{\sqrt{2\pi}} \int_{-\infty}^{r} e^{-\frac{u^2}{2}} \, du := \Phi(r)$$

that can not be calculated in a closed (explicit) form but rather is represented by widely available tables of values.

A random variable has a *Cauchy distribution* $P_X := \mathcal{C}(a, \sigma)$ with parameters $a \in \mathbb{R}, \sigma > 0$, if it has a density

$$p_X(r) = \frac{\sigma}{\pi(\sigma^2 + (r-a)^2)}, \qquad r \in \mathbb{R}.$$

The distribution $\mathcal{C}(0, 1)$ is called *standard Cauchy distribution*.

Exercise 1.4. Show how any Cauchy distribution can be reduced to or obtained from the standard Cauchy distribution by a linear transformation.

The *joint distribution* of random variables $X_1, \ldots, X_n$ defined on a common probability space is a measure $P_{X_1, \ldots, X_n}$ on $(\mathbb{R}^n, \mathcal{B}^n)$ given by

$$P_{X_1, \ldots, X_n}(A) = \mathbb{P}\{(X_1, \ldots, X_n) \in A\}, \qquad A \in \mathcal{B}^n.$$

The joint distribution determines the probabilities of all events depending only of the values $X_1, \ldots, X_n$, e.g. $\mathbb{P}(X_1 = X_2)$, $\mathbb{P}(X_n = \max_{1 \le j \le n} X_j)$ etc.

1.3 *Expectation*

Let X be a random variable on a probability space $(\Omega, \mathcal{A}, \mathbb{P})$. If the integral $\int_\Omega |X(\omega)|\mathbb{P}(d\omega)$ is finite, then we call *expectation* of X the value

$$\mathbb{E}\,X := \int_\Omega X(\omega)\mathbb{P}(d\omega).$$

In particular, if P_X is discrete, then

$$\mathbb{E}\,X := \sum_x x \cdot \mathbb{P}(X = x). \qquad (1.2)$$

Sometimes $\mathbb{E}\,X$ is called the *mean* of X.

The simplest properties of expectation are as follows.

- Expectation of a constant:

$$\mathbb{E}\,(c) = c.$$

- Linearity:

$$\mathbb{E}\,(X + Y) = \mathbb{E}\,X + \mathbb{E}\,Y; \qquad \mathbb{E}\,(cX) = c\,\mathbb{E}\,X. \qquad (1.3)$$

  Clearly, the summation rule extends to arbitrary finite number of summands. We have, as a special case, $\mathbb{E}\,(X + c) = \mathbb{E}\,X + c$.
- Monotonicity:

$$\mathbb{P}(X \geq 0) = 1 \rightsquigarrow \mathbb{E}\,X \geq 0.$$

  It follows that

$$\mathbb{P}(X \geq Y) = 1 \rightsquigarrow \mathbb{E}\,X \geq \mathbb{E}\,Y$$

  (if all expectations are well defined).
- Expectation through distribution. The most general formula we might need is as follows. Let $X_1, \ldots, X_n$ be random variables defined on a common probability space and let $f : \mathbb{R}^n \mapsto \mathbb{R}^1$ be a measurable function. Then

$$\mathbb{E}\,f(X_1, \ldots, X_n) = \int_{\mathbb{R}^n} f(x)P_{X_1,\ldots,X_n}(dx).$$

  In particular, for a single random variable $n = 1$ and we have

$$\mathbb{E}\,f(X) = \int_{\mathbb{R}^1} f(x)P_X(dx).$$

  By letting $f(x) = x$, we have

$$\mathbb{E}\,X = \int_{\mathbb{R}^1} x\,P_X(dx).$$

In particular, if P_X has a density $p(\cdot)$, we have

$$\mathbb{E}\,X = \int_{-\infty}^{\infty} x\,p(x)dx. \tag{1.4}$$

Notice that $\mathbb{E}\,X$ is well defined iff $\int_{\mathbb{R}^1} |x| P_X(dx) < \infty$ holds.
Let us stress that the existence and the value of $\mathbb{E}\,X$ are entirely determined by the distribution of X. In particular, identically distributed random variables have equal expectations.

- *Multiplication rule* for independent variables. Assume that random variables $X_1, \ldots, X_n$ are independent and their expectations are well defined. Then their product also has a well defined expectation and

$$\mathbb{E}\left(\prod_{j=1}^{n} X_j\right) = \prod_{j=1}^{n} (\mathbb{E}\,X_j). \tag{1.5}$$

The role of independence assumption becomes clear from the following example. Let a random variable X_1 takes the values ± 1 with equal probabilities $\frac{1}{2}$. Let $X_2 = X_1$. Then

$$\mathbb{E}\,(X_1 X_2) = \mathbb{E}\,X_1^2 = \mathbb{E}\,1 = 1,$$

while $\mathbb{E}\,(X_1)\mathbb{E}\,(X_2) = [\mathbb{E}\,X_1]^2 = 0$.
- A useful formula: if $X \geq 0$, then

$$\mathbb{E}\,X = \int_0^{\infty} \mathbb{P}(X \geq r)dr. \tag{1.6}$$

Proof.

$$\mathbb{E}\,X = \int_0^{\infty} x P_X(dx) = \int_0^{\infty} \int_0^{\infty} \mathbf{1}_{\{r \leq x\}} dr\, P_X(dx)$$

$$= \int_0^{\infty} dr \int_0^{\infty} \mathbf{1}_{\{x \geq r\}} P_X(dx)$$

$$= \int_0^{\infty} \mathbb{P}(X \geq r)dr.$$

$\square$

By dropping the assumption $X \geq 0$, we may obtain the following general result: if $\mathbb{E}\,|X| < \infty$, then

$$\mathbb{E}\,X = \int_0^{\infty} \mathbb{P}(X \geq r)dr - \int_{-\infty}^0 \mathbb{P}(X \leq r)dr.$$

The idea for proving this is to represent X as $X = X_+ - X_-$, where $X_+ = \max\{X, 0\}$, $X_- = \max\{-X, 0\}$, and apply (1.6) to X_+, X_-.

Here is couple of examples showing how to calculate expectations.

Example 1.5. Poisson distribution. Let $P_X = \mathcal{P}(a)$. Then by (1.2)

$$\mathbb{E}\,X = \sum_{k=0}^{\infty} k\,\mathbb{P}(X=k) = e^{-a}\sum_{k=0}^{\infty}\frac{ka^k}{k!} = e^{-a}\,a\sum_{k=1}^{\infty}\frac{a^{k-1}}{(k-1)!} = a. \qquad (1.7)$$

Example 1.6. Normal distribution. Let $P_X = \mathcal{N}(0,1)$. Then by (1.4)

$$\mathbb{E}\,X = \frac{1}{\sqrt{2\pi}}\int_{-\infty}^{\infty} xe^{-x^2/2}dx = 0.$$

By using the rules for linear transformations, (1.3) and (1.1), we easily extend this fact: if $P_X = \mathcal{N}(a,\sigma^2)$, then $\mathbb{E}\,X = a$.

Example 1.7. Cauchy distribution. Let $P_X = \mathcal{C}(0,1)$. Then the integral from (1.4)

$$\int_{-\infty}^{\infty}\frac{x\,dx}{\pi(1+x^2)}$$

diverges. Hence $\mathbb{E}\,X$ does not exist. Of course, the same is true for any Cauchy distribution.

1.4 *Inequalities based on expectation*

Let $f : \mathbb{R} \to \mathbb{R}_+$ be a non-decreasing function. Then

$$\mathbb{P}(X \geq r) \leq \frac{\mathbb{E}\,f(X)}{f(r)}, \qquad r \in \mathbb{R}. \qquad (1.8)$$

Proof. Compare two random variables: $Y_1 = f(X)$ and

$$Y_2 = \begin{cases} 0, & 0 \leq f(X) < r, \\ f(r), & r \leq f(X) < \infty. \end{cases}$$

Clearly, $Y_1 \geq Y_2$, whence

$$\mathbb{E}\,f(X) = \mathbb{E}\,Y_1 \geq \mathbb{E}\,Y_2 = f(r)\mathbb{P}(X \geq r). \qquad \square$$

Here are three useful examples of application of this inequality

1)

$$\mathbb{P}(|X| \geq r) \leq \frac{\mathbb{E}\,|X|}{r}, \qquad r \geq 0.$$

Here (1.8) is applied to the random variable $|X|$ and to the function

$$f(r) = \begin{cases} 0, & r < 0, \\ r, & r \geq 0. \end{cases}$$

2)

$$\mathbb{P}(|X| \geq r) \leq \frac{\mathbb{E}\,X^2}{r^2}, \qquad r \geq 0. \tag{1.9}$$

This inequality follows from the previous one by replacing X with X^2 and r with r^2.

3) *Exponential Chebyshev inequality*:

$$\mathbb{P}(X \geq r) \leq \frac{\mathbb{E}\,e^{\gamma X}}{e^{\gamma r}}, \qquad \gamma \geq 0,\ r \in \mathbb{R}.$$

We finish the list of inequalities related to expectations by stating *Jensen inequality*: Let $\mathbb{E}\,|X| < \infty$ and let a function $\varphi(\cdot)$ be convex. Then

$$\varphi(\mathbb{E}\,X) \leq \mathbb{E}\,\varphi(X).$$

The idea of the proof is as follows: if X takes a finite number of values: $\mathbb{P}(X = x_j) = p_j$, then by convexity

$$\varphi(\mathbb{E}\,X) = \varphi\left(\sum_j p_j x_j\right) \leq \sum_j p_j \varphi(x_j) = \mathbb{E}\,\varphi(X).$$

The general case follows by approximation arguments.

1.5 *Variance*

The *variance* of a random variable describes a measure of its dispersion around expectation. It is defined by

$$\mathbb{V}ar X := \mathbb{E}\,(X - \mathbb{E}\,X)^2.$$

The properties of variance are as follows.

- Alternative formula for variance:
$$\mathbb{V}ar X = \mathbb{E}\,(X^2) - (\mathbb{E}\,X)^2.$$

- The variance of a random variable is determined by its distribution,
$$\mathbb{V}ar X = \int_{\mathbb{R}} x^2 P_X(dx) - \left(\int_{\mathbb{R}} x P_X(dx)\right)^2.$$

In particular, identically distributed variables have equal variances.

- Positivity: $\mathbb{V}ar X \geq 0$.
- $\mathbb{V}ar(c) = 0$, $\mathbb{V}ar(X + c) = \mathbb{V}ar X$.
- 2-homogeneity:

$$\mathbb{V}ar(cX) = c^2 \, \mathbb{V}ar X, \qquad \forall c \in \mathbb{R}. \tag{1.10}$$

In particular, $\mathbb{V}ar(-X) = \mathbb{V}ar X$.

Accordingly, in order to get a 1-homogeneous measure of dispersion, one may consider the *standard deviation* defined by

$$\sigma(X) := \sqrt{\mathbb{V}ar X}.$$

Then, of course, $\sigma(cX) = |c|\sigma(X)$. Another advantage of standard deviation is that, along with expectation, it is "measured" in the same units as the random variable itself.

- *Addition rule.* The random variables X, Y satisfying assumptions

$$\mathbb{E}\, X^2 < \infty, \ \mathbb{E}\, Y^2 < \infty, \tag{1.11}$$

are called *uncorrelated,* if $\mathbb{E}\,(XY) = \mathbb{E}\, X \cdot \mathbb{E}\, Y$. It follows from (1.5) that independent random variables are uncorrelated, whenever assumption (1.11) holds.

If the random variables X, Y are uncorrelated, then we have addition rule

$$\mathbb{V}ar(X + Y) = \mathbb{V}ar X + \mathbb{V}ar Y.$$

The same assertion holds for the sum of any finite number of pairwise uncorrelated random variables.

In particular, we obtain a summation rule for the variances of independent variables: if the variables $X_1, \ldots, X_n$ are pairwise independent, $\mathbb{E}\, X_1^2 < \infty, \ldots, \mathbb{E}\, X_n^2 < \infty$, then

$$\mathbb{V}ar\left(\sum_{j=1}^{n} X_j\right) = \sum_{j=1}^{n} \mathbb{V}ar(X_j) \,.$$

Convenience of this summation rule gives to the variance a decisive advantage compared to other dispersion measures such as standard deviation or $\mathbb{E}\,|X - \mathbb{E}\, X|$.

- *Chebyshev inequality:*

$$\mathbb{P}\{|X - \mathbb{E}\, X| \geq r\} \leq \frac{\mathbb{V}ar X}{r^2}, \qquad \forall r > 0. \tag{1.12}$$

This inequality follows immediately from (1.9). It enables to evaluate the probability of the deviation of a random variable from its expectation.

- If $\mathbb{V}ar X = 0$, then there exists $c \in \mathbb{R}$ such that $\mathbb{P}(X = c) = 1$. This fact follows from Chebyshev inequality by letting $c = \mathbb{E}\,X$.

Here is couple of examples showing how to calculate variances.

Example 1.8. Poisson distribution. Let $P_X = \mathcal{P}(a)$. Then by going along the lines of Example 1.5 it is easy to see that $\mathbb{E}\,[X(X-1)] = a^2$. Hence, $\mathbb{E}\,X^2 = a^2 + \mathbb{E}\,X = a^2 + a$ and

$$\mathbb{V}ar X = (a^2 + a) - a^2 = a. \tag{1.13}$$

Example 1.9. Normal distribution. Let $P_X = \mathcal{N}(0,1)$. Then

$$\mathbb{V}ar X = \mathbb{E}\,X^2 = \int_{\mathbb{R}^1} x^2 P_X(dx)$$

$$= \frac{1}{\sqrt{2\pi}} \int_{-\infty}^{\infty} x^2 e^{-x^2/2} dx$$

$$= \frac{1}{\sqrt{2\pi}} \int_{-\infty}^{\infty} x \cdot x e^{-x^2/2} dx \qquad \text{(integrating by parts)}$$

$$= \frac{1}{\sqrt{2\pi}} \int_{-\infty}^{\infty} e^{-x^2/2} dx = 1.$$

By using the rules for linear transformations (1.10) and (1.1), we easily extend this fact: if $P_X = \mathcal{N}(a, \sigma^2)$, then $\mathbb{V}ar X = \sigma^2$.

1.6 *Covariance, correlation coefficient*

Covariance of two random variables X, Y is defined by

$$\text{cov}(X, Y) := \mathbb{E}\,((X - \mathbb{E}\,X)(Y - \mathbb{E}\,Y)) = \mathbb{E}\,(XY) - \mathbb{E}\,X\,\mathbb{E}\,Y,$$

if all expectations here are well defined. In particular, covariance is well defined when $\mathbb{V}ar X, \mathbb{V}ar Y < \infty$.

The trivial properties of covariance are as follows.

- Symmetry: $\text{cov}(X, Y) = \text{cov}(Y, X)$.
- Bilinearity:

$$\text{cov}(c\,X, Y) = c \cdot \text{cov}(X, Y),$$
$$\text{cov}(X_1 + X_2, Y) = \text{cov}(X_1, Y) + \text{cov}(X_2, Y).$$

- Connection to the variance: $\text{cov}(X, X) = \mathbb{V}ar X$.

- If the variables X, Y are uncorrelated, then $\text{cov}(X, Y) = 0$. In particular, if $\mathbb{E}X^2 < \infty, \mathbb{E}Y^2 < \infty$ and the variables X, Y are independent, then $\text{cov}(X, Y) = 0$.

Due to the latter fact, covariance is used for measuring the *linear* dependence between random variables. More precisely, the proper measure for linear dependence is a normalized quantity

$$\rho(X, Y) := \frac{\text{cov}(X, Y)}{\sqrt{\mathbb{V}arX}\,\sqrt{\mathbb{V}arY}},$$

called *correlation coefficient*.

The following example shows that correlation coefficient may not detect a *non-linear* dependence.

Example 1.10. Let X be a random variable such that $P_X = \mathcal{N}(0, 1)$ and let $Y := X^2$. The variables X and Y are clearly dependent but the covariance vanishes, since $\mathbb{E}X = 0$, $\mathbb{E}(XY) = \mathbb{E}X^3 = 0$. The same argument works for any random variable X having a symmetric distribution such that $\mathbb{E}|X|^3 < \infty$.

Recall that for random variables X, Y condition $\text{cov}(X, Y) = 0$ means that X and Y are uncorrelated. In case of non-zero covariance the random variables are called *positively correlated*, resp. *negatively correlated* whenever the covariance is positive, resp. negative. The properties of correlation coefficient are as follows.

- Normalization: for any X, Y it is true that

$$-1 \leq \rho(X, Y) \leq 1.$$

- 0-homogeneity:

$$\rho(cX + a, Y) = \begin{cases} \rho(X, Y), & c > 0, \\ -\rho(X, Y), & c < 0. \end{cases}$$

- Extremal values: for any X with $\mathbb{E}X^2 < \infty$ it is true that $\rho(X, X) = 1$, $\rho(X, -X) = -1$.
  Conversely, if $\rho(X, Y) = 1$, then X and Y differ only by a linear transformation, i.e. $Y = cX + a$, for some $a \in \mathbb{R}$ and $c > 0$. Similarly $\rho(X, Y) = -1$ yields $Y = cX + a$ with some $a \in \mathbb{R}$ and $c < 0$.

1.7 Complex-valued random variables

A *complex-valued random variable* X is a measurable mapping from a probability space to the complex plane, $X : (\Omega, \mathcal{A}, \mathbb{P}) \mapsto \mathbb{C}$. In other words, $X = X_1 + iX_2$, where X_1, X_2 is a pair of real-valued random variables.

The properties of complex-valued random variables are completely similar to those of the real ones. In particular,

$$\mathbb{E}\,X := \int_\Omega X\,d\mathbb{P} = \mathbb{E}\,X_1 + i\mathbb{E}\,X_2.$$

It is true that $\mathbb{E}\,(X + Y) = \mathbb{E}\,X + \mathbb{E}\,Y$, for $c \in \mathbb{C}$ we have $\mathbb{E}\,(cX) = c\,\mathbb{E}\,X$, and, whenever X, Y are independent, then $\mathbb{E}\,(XY) = \mathbb{E}\,X \cdot \mathbb{E}\,Y$. It is also true that $|\mathbb{E}\,X| \leq \mathbb{E}\,|X|$. For complex adjoint variables we additionally have $\mathbb{E}\,\overline{X} = \overline{\mathbb{E}\,X}$.

The only minor difference with the real case shows up in the definition of covariance, where now

$$\mathrm{cov}(X, Y) := \mathbb{E}\,\left((X - \mathbb{E}\,X)\overline{(Y - \mathbb{E}\,Y)} \right) = \mathbb{E}\,(X\overline{Y}) - \mathbb{E}\,X \cdot \overline{\mathbb{E}\,Y}.$$

With this definition

$$\mathrm{cov}(X, X) = \mathbb{E}\,|X - \mathbb{E}\,X|^2 \geq 0$$

is not called a variance anymore but still may be used as a measure of dispersion around expectation.

1.8 Characteristic functions

Characteristic function $f_X : \mathbb{R} \mapsto \mathbb{C}$ of a real-valued random variable X is defined by

$$f_X(t) := \mathbb{E}\,e^{itX}.$$

The simple properties of characteristic function are:

- $f_X(0) = 1$.
- $|f_X(t)| \leq 1$.
- $f_X(-t) = \overline{f_X(t)}$.
- $f_X(\cdot)$ is determined by the distribution of X, since

$$f_X(t) = \int_\mathbb{R} e^{itx} P_X(dx) = \int_\mathbb{R} \cos tx\, P_X(dx) + i \int_\mathbb{R} \sin tx\, P_X(dx).$$

In particular, if the distribution of X has a density p, then

$$f_X(t) = \int_\mathbb{R} e^{itx} p(x)dx.$$

Thus, up to a constant factor, characteristic function is the Fourier transform of the density p.

- If X and Y are independent, then $f_{X+Y}(t) = f_X(t)f_Y(t)$. In particular, if $X_1,\ldots,X_n$ are independent identically distributed (often abbreviated as *i.i.d.*) variables and $S := \sum_{1\leq j\leq n} X_j$, then $f_S(t) = f_{X_1}(t)^n$.
- Behavior by linear transformations: for any $a,c \in \mathbb{R}$ it is true that

$$f_{cX+a}(t) = e^{iat}f_X(ct). \tag{1.14}$$

- $f_X(\cdot)$ is a real function if and only if the distribution of X is symmetric, i.e. $P_X(B) = P_X(-B)$ for all $B \in \mathcal{B}^1$.
  Here the "if" statement is trivial, while "only if" statement is much less trivial. See Exercise 1.15 below.
- A function $f : \mathbb{R} \mapsto \mathbb{C}$ is called *non-negative definite*, if for all $n \in \mathbb{N}$, $t_1,\ldots,t_n \in \mathbb{R}$, $c_1,\ldots,c_n \in \mathbb{C}$

$$\sum_{j,k=1}^n f(t_j - t_k)c_j \overline{c_k} \geq 0. \tag{1.15}$$

Any characteristic function f_X is non-negative definite, since

$$\sum_{j,k=1}^n f(t_j - t_k)c_j \overline{c_k} = \sum_{j,k=1}^n \mathbb{E}\, e^{i(t_j-t_k)X}c_j\overline{c_k}$$

$$= \mathbb{E}\left(\sum_{j,k=1}^n e^{it_jX}c_j\overline{e^{it_kX}\overline{c_k}}\right)$$

$$= \mathbb{E}\left(\sum_{j=1}^n e^{it_jX}c_j\overline{\sum_{k=1}^n e^{it_kX}c_k}\right)$$

$$= \mathbb{E}\left|\sum_{j=1}^n e^{it_jX}c_j\right|^2 \geq 0.$$

Actually, the famous *Bochner–Khinchin theorem* asserts that if $f : \mathbb{R} \mapsto \mathbb{C}$ is non-negatively defined, continuous and $f(0) = 1$, then there exists a random variable X such that $f_X = f$.

Examples of characteristic functions:

Example 1.11. Poisson distribution. Let $P_X = \mathcal{P}(a)$. Then

$$f_X(t) = \sum_{k=0}^\infty \mathbb{P}(X=k)e^{itk} = e^{-a}\sum_{k=0}^\infty \frac{a^k}{k!}\, e^{itk} = e^{-a+ae^{it}} = e^{a(e^{it}-1)}. \tag{1.16}$$

Example 1.12. Normal distribution. Let $P_X = \mathcal{N}(0,1)$. Then

$$f_X(t) = \frac{1}{\sqrt{2\pi}} \int_{-\infty}^{\infty} e^{itu} e^{-u^2/2} du = \frac{e^{-t^2/2}}{\sqrt{2\pi}} \int_{-\infty}^{\infty} e^{-(u-it)^2/2} du$$

$$= \frac{e^{-t^2/2}}{\sqrt{2\pi}} \int_{-\infty+it}^{\infty+it} e^{-(u-it)^2/2} du = \frac{e^{-t^2/2}}{\sqrt{2\pi}} \int_{-\infty}^{\infty} e^{-v^2/2} dv = e^{-t^2/2}.$$

By the linear transformation rules (1.1), (1.14), whenever $P_X = \mathcal{N}(a,\sigma^2)$, we have

$$f_X(t) = e^{iat-\sigma^2 t^2/2}. \tag{1.17}$$

Example 1.13. Cauchy distribution. Let $P_X = \mathcal{C}(0,1)$. Then by using the well known expression for a definite integral,

$$f_X(t) = \frac{1}{\pi} \int_{-\infty}^{\infty} \frac{e^{itu} du}{1+u^2} = \frac{2}{\pi} \int_{0}^{\infty} \frac{\cos(tu) du}{1+u^2} = e^{-|t|}.$$

By the linear transformation rules (1.1), (1.14), whenever $P_X = \mathcal{C}(a,\sigma)$, we have

$$f_X(t) = e^{iat-\sigma|t|}. \tag{1.18}$$

The key point is the reconstruction of distribution from the corresponding characteristic function. For example, if the characteristic function f_X is integrable, then the distribution P_X has a density $p(\cdot)$ and we can reconstruct this density by inverse Fourier transform

$$p(x) = \frac{1}{2\pi} \int_{\mathbb{R}} e^{-itx} f_X(t) dt.$$

However, in general, the reconstruction procedure, given by the following inversion formula, is much more sophisticated.

Theorem 1.14 (Inversion formula). *If the numbers $a,b \in \mathbb{R}$ are such that $a < b$ and $\mathbb{P}(X=a) = \mathbb{P}(X=b) = 0$, then*

$$P_X[a,b] = \frac{1}{2\pi} \lim_{T\to\infty} \int_{-T}^{T} \frac{e^{-ita} - e^{-itb}}{it} f_X(t) dt. \tag{1.19}$$

The precise form of the inversion formula is unimportant for us. We must only retain the fact that the distribution of a random variable (the left hand side of (1.19)) is completely determined by the characteristic function (the right hand side of (1.19)).

Exercise 1.15. By using inversion formula, prove that if $f_X(\cdot)$ is real, then the distribution P_X is symmetric.

1.9 *Convergence of random variables*

As in Calculus, the notions of limit and convergence play in Probability the basic role and serve for description of the main probabilistic distributions. The things are, however, complicated by the fact that there are *several* types of convergence of random variables.

Let X be a random variable and let (X_n) be a sequence of random variables. Unless opposite is stated explicitly, we assume that they are defined on a common probability space $(\Omega, \mathcal{A}, \mathbb{P})$. Let us consider four basic types of convergence of random variables.

- **Almost sure convergence** (a.s. convergence or convergence with probability one). $X_n \xrightarrow{\text{a.s.}} X$, iff

$$\mathbb{P}\{\omega \in \Omega : \lim_{n \to \infty} X_n(\omega) = X(\omega)\} = 1.$$

- **Convergence in the mean** (of order $p > 0$). $X_n \xrightarrow{\mathbb{L}_p} X$, iff

$$\lim_{n \to \infty} \mathbb{E} |X_n - X|^p = 0.$$

- **Convergence in probability.** $X_n \xrightarrow{\mathbb{P}} X$, iff

$$\lim_{n \to \infty} \mathbb{P}\{|X_n - X| \geq \varepsilon\} = 0, \qquad \forall \varepsilon > 0.$$

- **Convergence in distribution** (convergence in law, or weak convergence). $X_n \Rightarrow X$, iff the distribution functions converge,

$$\lim_{n \to \infty} F_{X_n}(r) = F_X(r), \qquad \forall r \in C_X,$$

where C_X is the set of continuity points of the limiting distribution function $F_X(\cdot)$.

  The latter is the only type of convergence that does not require from the variables X_n to be defined on a common probability space.

Remark 1.16. If all random variables X_n are degenerated, which means that $\mathbb{P}(X_n = c_n) = 1$, and $c_n \searrow c$, then $X_n \Rightarrow X$, but on the other hand we have $0 = F_{X_n}(c) \nrightarrow 1 = F_X(c)$. Therefore, we can not impose convergence $F_{X_n}(r) \to F_X(r)$ *for all* $r \in \mathbb{R}$ in the definition of convergence in distribution.

In many important cases one has $X_n \Rightarrow X$ with X having a normal distribution. Then one says that the sequence (X_n) obeys *central limit theorem*.

Figure 1.1 illustrates the connection between different types of convergence.

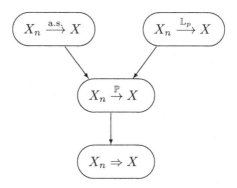

Fig. 1.1   Relations between convergence types.

Remark 1.17. Convergence in probability does not imply almost sure convergence. However, from any sequence converging in probability one can extract a *subsequence* converging almost surely.

Remark 1.18. Convergence in distribution does not imply convergence in probability. However, these types of convergence are equivalent if the limiting random variable is a constant.

Indeed, let $X_n \Rightarrow X$ and assume that X has a degenerated distribution, i.e. for some $c \in \mathbb{R}$ it is true that $\mathbb{P}(X = c) = 1$. Then $C_X = \mathbb{R}\backslash\{c\}$ and convergence in distribution means that

$$\lim_{n \to \infty} F_{X_n}(r) = \begin{cases} 0, & r < c, \\ 1, & r > c. \end{cases}$$

By letting $r = c \pm h$, we have

$$\lim_{n \to \infty} \mathbb{P}(|X_n - c| > h)$$
$$\leq \lim_{n \to \infty} \mathbb{P}(X_n > c + h) + \lim_{n \to \infty} \mathbb{P}(X_n \leq c - h)$$
$$= \lim_{n \to \infty} [1 - F_{X_n}(c + h)] + \lim_{n \to \infty} F_{X_n}(c - h) = 0.$$

Hence, $X_n \xrightarrow{\mathbb{P}} X$.

Exercise 1.19. Let (X_n) and (Y_n) be two sequences of random variables on a common probability space such that $X_n \Rightarrow X$ and $Y_n \xrightarrow{\mathbb{P}} 0$. Prove that $X_n + Y_n \Rightarrow X$.

There is a useful *Ky Fan distance* between the random variables related to convergence in probability. For two random variables X, Y defined on a common probability space let

$$d_{KF}(X,Y) := \inf\{\varepsilon : \mathbb{P}(|X - Y| > \varepsilon) < \varepsilon\}.$$

Exercise 1.20. Prove the triangle inequality for Ky Fan distance: for any random variables X, Y, Z defined on a common probability space it is true that

$$d_{KF}(X,Z) \leq d_{KF}(X,Y) + d_{KF}(Y,Z).$$

Remarkably, we have $X_n \xrightarrow{\text{P}} X$ iff $d_{KF}(X_n, X) \to 0$. Moreover, the class of all random variables equipped with the distance $d_{KF}(\cdot, \cdot)$ is a complete metric space. This means that if we have a Cauchy sequence (X_n) of random variables, i.e.

$$\lim_{n \to \infty} \sup_{n_1, n_2 \geq n} d_{KF}(X_{n_1}, X_{n_2}) = 0, \tag{1.20}$$

then (X_n) converges in probability to some random variable.

There exist a number of equivalent ways to define convergence in distribution. We enlist them in the following result.

Theorem 1.21. *Let (X_n) and X be some random variables. Then the following assertions are equivalent.*
a) $X_n \Rightarrow X$.
b) For any closed set $B \subset \mathbb{R}$ it is true that

$$\limsup_{n \to \infty} \mathbb{P}\{X_n \in B\} \leq \mathbb{P}\{X \in B\}.$$

c) For any open set $V \subset \mathbb{R}$ it is true that

$$\liminf_{n \to \infty} \mathbb{P}\{X_n \in V\} \geq \mathbb{P}\{X \in V\}.$$

d) For any measurable set $A \subset \mathbb{R}$, satisfying regularity condition with respect to X, i.e. $\mathbb{P}(X \in \partial A) = 0$, it is true that

$$\lim_{n \to \infty} \mathbb{P}\{X_n \in A\} = \mathbb{P}\{X \in A\}.$$

e) For any bounded continuous function $f : \mathbb{R} \mapsto \mathbb{R}$ it is true that

$$\lim_{n \to \infty} \mathbb{E} f(X_n) = \mathbb{E} f(X).$$

Finally, let us express convergence in distribution in terms of characteristic functions. First, if $X_n \Rightarrow X$, then by Theorem 1.21 e) we have the convergence of characteristic functions:

$$\lim_{n \to \infty} f_{X_n}(t) \to f_X(t), \qquad \forall t \in \mathbb{R}.$$

It is much more important to invert this statement: be able to derive the convergence in distribution from the convergence of characteristic functions. Here the following is true.

Theorem 1.22. *Let (X_n) be a sequence of random variables such that*

$$\lim_{n \to \infty} f_{X_n}(t) \to f(t), \qquad \forall t \in \mathbb{R},$$

while the limit function $f(\cdot)$ is continuous at zero. Then there exists a random variable X such that

$$X_n \Rightarrow X,$$

and we have $f_X(\cdot) = f(\cdot)$.

Notice that this result is particularly convenient. First, we need not prove that the limit function is a characteristic function – this comes for free. Second, we may consider each argument value t separately because no uniform convergence is assumed.

2 From Poisson to Stable Variables

2.1 *Compound Poisson variables*

We will use a customary notation δ_u for the unit measure concentrated at a point u, i.e.

$$\delta_u(B) := \begin{cases} 1, & u \in B; \\ 0, & u \notin B. \end{cases}$$

Let positive $a_1, \ldots, a_J$ and non-zero real $u_1, \ldots, u_J$ be given. Next, let X_j, $1 \le j \le J$, be independent Poisson random variables such that $P_{X_j} = \mathcal{P}(a_j)$. Let $S := \sum_{j=1}^{J} u_j X_j$. We say that S has a *compound Poisson distribution* with finitely supported *intensity measure* $\nu := \sum_{j=1}^{J} a_j \delta_{u_j}$, which means that

$$\nu(B) = \sum_{j=1}^{J} a_j \mathbf{1}_{\{u_j \in B\}}, \qquad B \in \mathcal{B}.$$

All characteristics of S can be easily expressed via its intensity measure ν:

$$\mathbb{E}\, S = \sum_{j=1}^{J} u_j \mathbb{E}\, X_j = \sum_{j=1}^{J} u_j a_j = \int u\,\nu(du),$$

$$\mathbb{V}ar S = \sum_{j=1}^{J} u_j^2 \mathbb{V}ar X_j = \sum_{j=1}^{J} u_j^2 a_j = \int u^2 \nu(du),$$

$$\mathbb{E}\, e^{itS} = \prod_{j=1}^{J} \mathbb{E}\, e^{itu_j X_j} = \prod_{j=1}^{J} \exp\left\{ a_j(e^{itu_j} - 1) \right\}$$

$$= \exp\left\{ \int (e^{itu} - 1)\nu(du) \right\}.$$

We may also consider a *centered* compound Poisson random variable with intensity measure ν, namely $\overline{S} := S - \mathbb{E}\, S$. Clearly, we have

$$\mathbb{E}\, \overline{S} = 0, \tag{2.1}$$

$$\mathbb{V}ar \overline{S} = \mathbb{V}ar S = \int u^2 \nu(du), \tag{2.2}$$

$$\mathbb{E}\, e^{it\overline{S}} = e^{-it\mathbb{E}\, S}\, \mathbb{E}\, e^{itS} = \exp\left\{ \int (e^{itu} - 1 - itu)\nu(du) \right\}. \tag{2.3}$$

Our next move will be to define a compound Poisson random variable with arbitrary (not necessarily finitely supported) intensity. Let ν be an arbitrary non-zero finite measure on $\mathbb{R}\backslash\{0\}$. We wish to construct a random variable S such that, as above,

$$\mathbb{E}\, e^{itS} = \exp\left\{ \int (e^{itu} - 1)\nu(du) \right\}. \tag{2.4}$$

Informally, if ν is a continuous measure, one should percept the corresponding compound Poisson random variable as an infinite sum of infinitely small independent Poisson random variables. The precise construction is, however, as follows.

Let $|\nu| := \nu(\mathbb{R}) > 0$ be the total mass of ν. Then $\frac{\nu}{|\nu|}$ is a probability distribution, and we may consider a sequence of i.i.d. random variables U_j such that $P_{U_j} = \frac{\nu}{|\nu|}$. Furthermore, let N be a Poisson random variable with distribution $P_N = \mathcal{P}(|\nu|)$, independent of (U_j). Consider a sum of *random number* of U_j's:

$$S := \sum_{j=1}^{N} U_j, \tag{2.5}$$

where we formally let $S = 0$ whenever $N = 0$. Then we have

$$\mathbb{E}\, e^{itS} = \sum_{k=0}^{\infty} \mathbb{P}(N = k)\, \mathbb{E}\, \exp\left\{it \sum_{j=1}^{k} U_j\right\}$$

$$= \sum_{k=0}^{\infty} \frac{e^{-|\nu|}|\nu|^k}{k!}\, (\mathbb{E}\, \exp\{itU_1\})^k$$

$$= e^{-|\nu|} \exp\left\{|\nu|\mathbb{E}\, \exp\{itU_1\}\right\}$$

$$= \exp\left\{-|\nu| + |\nu| \int e^{itu}\frac{\nu(du)}{|\nu|}\right\}$$

$$= \exp\left\{\int (e^{itu} - 1)\nu(du)\right\}, \tag{2.6}$$

as required.

Under appropriate assumptions on ν we still have the former expressions for expectation and variance. Namely, if

$$\int |u|\nu(du) = |\nu|\,\mathbb{E}\, U_j < \infty,$$

then

$$\mathbb{E}\, S = \sum_{k=0}^{\infty} \mathbb{P}(N = k)\, \mathbb{E}\, \Big(\sum_{j=1}^{k} U_j\Big)$$

$$= \sum_{k=0}^{\infty} \mathbb{P}(N = k)\, k\,\mathbb{E}\, U_1$$

$$= \mathbb{E}\, N \cdot \mathbb{E}\, U_1$$

$$= |\nu| \cdot \mathbb{E}\, U_1 = \int u\nu(du).$$

Similarly, if

$$\int u^2\nu(du) = |\nu|\,\mathbb{E}\, U_j^2 < \infty,$$

then

$$\mathbb{E}\, S^2 = \sum_{k=0}^{\infty} \mathbb{P}(N=k)\, \mathbb{E}\left(\sum_{j=1}^{k} U_j\right)^2$$

$$= \sum_{k=0}^{\infty} \mathbb{P}(N=k)\left[\mathbb{V}ar\left(\sum_{j=1}^{k} U_j\right) + \left(\mathbb{E}\left(\sum_{j=1}^{k} U_j\right)\right)^2\right]$$

$$= \sum_{k=0}^{\infty} \mathbb{P}(N=k)\left[k\,\mathbb{V}ar U_1 + (k\mathbb{E}\, U_1)^2\right]$$

$$= \mathbb{E}\, N \cdot \mathbb{V}ar U_1 + \mathbb{E}\, N^2 \cdot (\mathbb{E}\, U_1)^2$$

$$= \mathbb{E}\, N \cdot (\mathbb{E}\, U_1^2 - (\mathbb{E}\, U_1)^2) + \mathbb{E}\, N^2 \cdot (\mathbb{E}\, U_1)^2$$

$$= \mathbb{E}\, N \cdot \mathbb{E}\, U_1^2 + (-\mathbb{E}\, N + \mathbb{V}ar N + (\mathbb{E}\, N)^2) \cdot (\mathbb{E}\, U_1)^2$$

$$= |\nu| \cdot \mathbb{E}\, U_1^2 + (-|\nu| + |\nu| + |\nu|^2)(\mathbb{E}\, U_1)^2$$

$$= |\nu| \cdot \mathbb{E}\, U_1^2 + |\nu|^2 (\mathbb{E}\, U_1)^2 \,.$$

Hence,

$$\mathbb{V}ar S = \mathbb{E}\, S^2 - (\mathbb{E}\, S)^2 = |\nu|\, \mathbb{E}\, (U_1^2) = \int u^2 \nu(du).$$

For the centered compound Poisson variable $\overline{S} := S - \mathbb{E}\, S$ all formulae (2.1), (2.2), (2.3) also remain valid.

Exercise 2.1. Prove that

$$\mathbb{E}\, \overline{S}^4 = \int u^4 \nu(du) + 3\left(\int u^2 \nu(du)\right)^4. \qquad (2.7)$$

2.2   *Limits of compound Poisson variables*

We proceed now to further extending the class of compound Poisson random variables by handling the appropriate classes of *infinite* intensity measures. The corresponding distributions are obtained as the limits of compound Poisson distributions (non-centered, centered, or partially centered[1]).

Let ν be a non-zero finite measure on $\mathbb{R}\backslash\{0\}$. Assume that

$$\int \min\{|u|, 1\}\nu(du) < \infty. \qquad (2.8)$$

---

[1]The precise meaning of partial centering is explained below.

Notice that any finite measure satisfies this condition but there are also infinite measures (accumulating infinite mass at zero) satisfying it. We say that the distribution of a random variable S belongs to a class $\mathcal{P}_{1,0}$ and write $S \in \mathcal{P}_{1,0}$, if its characteristic function writes as

$$\mathbb{E}\, e^{itS} = \exp\left\{ \int (e^{itu} - 1)\nu(du) \right\} \qquad (2.9)$$

for some measure ν satisfying (2.8). Notice that for any real t the integral on the right hand side is well defined, since $|e^{itu} - 1| \le \min\{|t|\,|u|, 2\}$.

The subscripts one and zero in the notation $\mathcal{P}_{1,0}$ indicate the degrees of u at zero and at infinity involved in formula (2.8). We will also need classes $\mathcal{P}_{2,1}$ and $\mathcal{P}_{2,0}$ later on.

In the following we will show that the integrals over Poisson random measure with independent values belong to $\mathcal{P}_{1,0}$.

The next proposition shows that the distributions from $\mathcal{P}_{1,0}$ do exist.

Proposition 2.2. *Let ν be a measure satisfying (2.8). Let ν_n be the restriction of ν on $\mathbb{R}\setminus[-\frac{1}{n}, \frac{1}{n}]$. Let S_n be the compound Poisson random variables with intensity ν_n. Then S_n converge in distribution to a random variable S with characteristic function (2.9).*

Proof. Notice first of all that ν_n is a finite measure, hence S_n is well defined. Indeed,

$$\nu_n(\mathbb{R}) = \nu\left(\mathbb{R}\setminus[-\tfrac{1}{n}, \tfrac{1}{n}]\right) = \int_{\mathbb{R}\setminus[-\frac{1}{n},\frac{1}{n}]} 1\,\nu(du)$$

$$\le \int_{\mathbb{R}\setminus[-\frac{1}{n},\frac{1}{n}]} n \min\{|u|, 1\}\,\nu(du) \le n \int_{\mathbb{R}} \min\{|u|, 1\}\,\nu(du) < \infty.$$

We will use the convergence criterion in terms of characteristic functions, Theorem 1.22.

First we have to show that for any fixed $t \in \mathbb{R}$ the characteristic functions $\mathbb{E}\, e^{itS_n}$ converge to (2.9). Since by (2.4)

$$\mathbb{E}\, e^{itS_n} = \exp\left\{ \int (e^{itu} - 1)\nu_n(du) \right\} = \exp\left\{ \int_{\mathbb{R}\setminus[-\frac{1}{n},\frac{1}{n}]} (e^{itu} - 1)\nu(du) \right\},$$

it remains to prove that

$$\lim_{n\to\infty} \int_{\mathbb{R}\setminus[-\frac{1}{n},\frac{1}{n}]} (e^{itu} - 1)\nu(du) = \int_{\mathbb{R}} (e^{itu} - 1)\nu(du),$$

which follows from

$$\left| \int_{\mathbb{R}} (e^{itu} - 1)\nu(du) - \int_{\mathbb{R} \setminus [-\frac{1}{n}, \frac{1}{n}]} (e^{itu} - 1)\nu(du) \right|$$

$$= \left| \int_{[-\frac{1}{n}, \frac{1}{n}]} (e^{itu} - 1)\nu(du) \right|$$

$$\leq \int_{[-\frac{1}{n}, \frac{1}{n}]} \left| e^{itu} - 1 \right| \nu(du) \leq |t| \int_{[-\frac{1}{n}, \frac{1}{n}]} |u|\nu(du) \to 0, \qquad \text{as } n \to 0.$$

Notice that here we used (2.8) at the last step.

Finally, in order to apply Theorem 1.22, we have to check that the limit function in (2.9) is continuous at zero. In other words, we must prove that

$$\lim_{t \to 0} \int (e^{itu} - 1)\nu(du) = 0.$$

Indeed, we have

$$\left| \int (e^{itu} - 1)\nu(du) \right| \leq \int \left| e^{itu} - 1 \right| \nu(du)$$

$$\leq \int \min\{|t|\,|u|, 2\}\nu(du) \to 0,$$

by Lebesgue dominated convergence theorem. The application of the latter is justified by (2.8).

Now Theorem 1.22 applies and provides the required convergence in distribution of S_n. $\qquad \square$

The expressions for $\mathbb{E}\,S$ and $\mathbb{V}ar X$ through intensity obtained for compound Poisson variables remain true for $S \in \mathcal{P}_{1,0}$. Namely, if it is true that $\int |u|\nu(du) < \infty$, then

$$\mathbb{E}\,S = \int u\nu(du);$$

while if $\int u^2\nu(du) < \infty$, then

$$\mathbb{V}ar S = \int u^2\nu(du).$$

Next, let us see which distributions emerge as the limits of *centered* compound Poisson distributions. Let ν be a non-zero finite measure on $\mathbb{R} \setminus \{0\}$. Assume that

$$\int \min\{|u|^2, |u|\}\nu(du) < \infty. \qquad (2.10)$$

Again there exist infinite measures (accumulating infinite mass at zero) satisfying this condition.

We say that the distribution of a random variable $\overline{S}$ belongs to a class $\mathcal{P}_{2,1}$ and write $\overline{S} \in \mathcal{P}_{2,1}$, if its characteristic function writes as

$$\mathbb{E}\, e^{it\overline{S}} = \exp\left\{ \int (e^{itu} - 1 - itu)\nu(du) \right\} \qquad (2.11)$$

for some measure ν satisfying (2.10). Notice that for any real t the integral on the right hand side is well defined, since

$$|e^{itu} - 1 - itu| \leq \min\{\tfrac{t^2u^2}{2}, 2 + |t|\,|u|\} \leq \max\{\tfrac{t^2}{2}, 2 + |t|\} \cdot \min\{|u|^2, |u|\}.$$

In the following we will show that the integrals over *centered* Poisson random measure with independent values belong to $\mathcal{P}_{2,1}$.

The next proposition shows that the distributions from $\mathcal{P}_{2,1}$ do exist.

Proposition 2.3. *Let ν be a measure satisfying (2.10). Let ν_n be the restriction of ν on $\mathbb{R}\setminus[-\tfrac{1}{n}, \tfrac{1}{n}]$. Let $\overline{S}_n$ be the centered compound Poisson random variables with intensity ν_n. Then $\overline{S}_n$ converge in distribution to a random variable S with characteristic function (2.11).*

The proof of Proposition (2.3) follows the same lines as that of Proposition (2.2). We leave it to the reader as an exercise.

The expressions for $\mathbb{E}\,\overline{S}$ and $\mathbb{V}ar\overline{S}$ through intensity obtained for centered compound Poisson variables remain true for $\overline{S} \in \mathcal{P}_{2,1}$. Namely, if $\int |u|\nu(du) < \infty$, then $\mathbb{E}\,\overline{S} = 0$, while if $\int u^2\nu(du) < \infty$, then

$$\mathbb{V}ar\overline{S} = \int u^2\nu(du).$$

Notice that the classes of measures satisfying (2.8) and (2.10) are not comparable: it is fairly easy to construct a measure ν that satisfies any of these conditions while the other one fails.

Surprisingly, the largest class of intensities, including both those of $\mathcal{P}_{1,0}$ and $\mathcal{P}_{2,1}$ is achieved by *partial centering* of compound Poisson variables.

Let us split the real line $\mathbb{R}$ into two parts: $\mathcal{I} := [-1, 1]$ and

$$\mathcal{O} := \mathbb{R}\setminus\mathcal{I} = \{l : |l| > 1\}.$$

This splitting is completely arbitrary but it is necessary for us to act in special way at *some* neighborhood of zero.

Let ν be a non-zero finite measure on $\mathbb{R}\backslash\{0\}$. Assume that *Lévy–Khinchin condition*

$$\int \min\{|u|^2, 1\}\nu(du) < \infty \tag{2.12}$$

holds. Let $\nu_{\mathcal{I}}$ and $\nu_{\mathcal{O}}$ be the restrictions of ν on $\mathcal{I}$ and $\mathcal{O}$, respectively. Then $\nu_{\mathcal{I}}$ has the finite second moment and satisfies (2.10), while $\nu_{\mathcal{O}}$ is a finite measure. Therefore a variable $\overline{S}_{\mathcal{I}}$, corresponding to $\nu_{\mathcal{I}}$ and a compound Poisson random variable $S_{\mathcal{O}}$ are well defined. Moreover, assume that $\overline{S}_{\mathcal{I}}$ and $S_{\mathcal{O}}$ are independent and let $\widetilde{S} := \overline{S}_{\mathcal{I}} + S_{\mathcal{O}}$. Then

$$\mathbb{E}\,e^{it\widetilde{S}} = \exp\left\{\int (e^{itu} - 1 - itu\,\mathbf{1}_{\{|u|\leq 1\}})\nu(du)\right\}. \tag{2.13}$$

We say that the distribution of a random variable $\widetilde{S}$ belongs to a class $\mathcal{P}_{2,0}$ and write $\widetilde{S} \in \mathcal{P}_{2,0}$, if its characteristic function writes as in (2.13). for some measure ν satisfying (2.12).

Notice that Lévy-Khinchin condition is weaker that any of conditions (2.8), (2.10). Hence it applies to a wider class of intensity measures ν.

As for the expectation and variance of $\widetilde{S} \in \mathcal{P}_{2,0}$, it follows from the previous results that if $\int_{\mathcal{O}} |u|\nu(du) < \infty$, then

$$\mathbb{E}\,\widetilde{S} = \int_{\mathcal{O}} u\nu(du),$$

while if $\int u^2\nu(du) < \infty$, then

$$\mathbb{V}ar\widetilde{S} = \int u^2\nu(du).$$

2.3   *A mystery at zero*

A careful reader would remark that, strangely enough, we do not allow the intensity measure to charge the point zero. Formally, it is clear from the very first definition of compound Poisson variable with finitely supported intensity measure $S := \sum_{j=1}^J u_j X_j$, that if we allow some u_j be equal to zero, one may drop the term $u_j X_j$ without any affect on S. Therefore, adding intensity charges at zero is useless.

There is, however, a deeper thing one should understand about intensity at zero: the normal distribution is hidden there! This means that if some intensity measures ν_n converge to zero in some sense while their second moments remain constant: $\sigma^2 = \int u^2\nu_n(du)$ then the distributions of the corresponding centered compound Poisson random variables converge to the normal distribution $\mathcal{N}(0, \sigma^2)$. Let us illustrate this fact by a quick example. Let

$$\nu_n := \frac{n^2}{2} \left(\delta_{-1/n} + \delta_{1/n} \right),$$

a measure placing equal large weights at the points $\pm \frac{1}{n}$ going to zero. Then the corresponding compound Poisson random variables S_n (which are automatically centered because of the symmetry of ν_n) have characteristic functions

$$\mathbb{E}\, e^{itS_n} = \exp \left\{ \int (e^{itu} - 1)\nu_n(du) \right\}$$

$$= \exp \left\{ n^2 (e^{-it/n} + e^{it/n} - 2)/2 \right\} = \exp \left\{ -n^2(1 - \cos(t/n)) \right\}$$

$$= \exp \left\{ -2n^2 \sin(t/2n)^2 \right\} \to \exp \left\{ -t^2/2 \right\}, \qquad \text{as } n \to \infty.$$

Since the limit is the characteristic function of $\mathcal{N}(0, 1)$, the convergence of $\mathbb{E}\, e^{itS_n}$ exactly means the convergence of S_n in distribution to this normal distribution.

2.4 *Infinitely divisible random variables*

A random variable Y is called *infinitely divisible* if for each $n = 1, 2, \ldots$ there exist i.i.d. random variables $X_1, \ldots, X_n$ such that the distribution of Y coincides with the distribution of $\sum_{j=1}^{n} X_j$. In the language of characteristic functions this means that for each $n = 1, 2, \ldots$ there exists a characteristic function f_n such that

$$f_Y(t) = f_n(t)^n, \qquad t \in \mathbb{R}. \tag{2.14}$$

The following theorem essentially shows that any infinitely divisible random variable can be obtained by summation of a normal random variable with independent random variable from Poisson class $\mathcal{P}_{2,0}$.

Theorem 2.4 (Lévy–Khinchin representation). *A random variable Y is infinitely divisible iff its characteristic function can be written in the form*

$$\mathbb{E}\, e^{itY} = \exp \left\{ iat - \frac{\sigma^2 t^2}{2} + \int (e^{itu} - 1 - itu\, \mathbf{1}_{\{|u| \le 1\}})\nu(du) \right\} \tag{2.15}$$

with some unique real a, $\sigma \ge 0$ and a measure ν on $\mathbb{R} \backslash \{0\}$ satisfying integrability condition (2.12).

It is trivial to prove that any function

$$f(t) := \exp\left\{ iat - \frac{\sigma^2}{2} + \int (e^{itu} - 1 - itu\,\mathbf{1}_{\{|u|\leq 1\}})\nu(du) \right\}$$

with a, σ and ν as in theorem is a characteristic function and that it corresponds to an infinitely divisible random variable. Indeed, let Y_1 be an $\mathcal{N}(a, \sigma^2)$ random variable and let $Y_2 \in \mathcal{P}_{2,0}$ be a random variable with characteristic function (2.13) (we know now, after all efforts related to compound Poisson variables, that Y_2 exists). Assuming that Y_1 and Y_2 are independent, f is a characteristic function of $Y := Y_1 + Y_2$. Moreover, for each n let $f_n(t) := f_{n,1}(t)f_{n,2}(t)$ where $f_{n,1}(t) := \exp\{iat/n - \sigma^2 t^2/2n\}$ is a characteristic function of the normal distribution $\mathcal{N}(\frac{a}{n}, \frac{\sigma^2}{n})$, and

$$f_{n,2}(t) := \exp\left\{ \int (e^{itu} - 1 - itu\,\mathbf{1}_{\{|u|\leq 1\}})\frac{\nu(du)}{n} \right\}$$

is a characteristic function from the class $\mathcal{P}_{2,0}$ with intensity $\frac{\nu}{n}$. With this choice of f, we immediately obtain (2.14).

The converse claim of Theorem 2.4, asserting that any infinitely divisible random variable has a characteristic function (2.15) is much harder to prove. However, since we will never use this fact, we do not present its proof here.

One should retain from Theorem 2.4 that an infinitely divisible distribution is characterized by the *triplet* (a, σ^2, ν). Moreover, there is an obvious summation rule for triplets: if Y_1 and Y_2 are independent infinitely divisible variables characterized by their respective triplets (a_1, σ_1^2, ν_1), (a_2, σ_2^2, ν_2), then $Y_1 + Y_2$ also is infinitely divisible and corresponds to the triplet $(a_1 + a_2, \sigma_1^2 + \sigma_2^2, \nu_1 + \nu_2)$.

2.5   *Stable variables*

We will now consider one special subclass of infinitely divisible variables and distributions that plays particularly important role in probabilistic limit theorems.

A random variable X (and its distribution) is called *stable*, if for any independent variables X_1 and X_2 equidistributed with X and any $k_1, k_2 > 0$ there exist $k_3 > 0$ and $k_4 \in \mathbb{R}$ such that $k_1 X_1 + k_2 X_2$ is equidistributed with $k_3 X + k_4$. Moreover, X is called *strictly stable*, if $k_4 = 0$, i.e.   for any $k_1, k_2 > 0$ there exists $k_3 > 0$ such that $k_1 X_1 + k_2 X_2$ is equidistributed with $k_3 X$.

As an example, let us show that all normal random variables are stable: Notice first that for any independent X_1, X_2 normal random variables with distributions $P_{X_1} = \mathcal{N}(a_1, \sigma_1^2)$ and $P_{X_2} = \mathcal{N}(a_2, \sigma_2^2)$ the sum $X_1 + X_2$ is also normal, namely

$$P_{X_1+X_2} = \mathcal{N}(a_1 + a_2, \sigma_1^2 + \sigma_2^2). \tag{2.16}$$

We can check this by computing the characteristic function

$$f_{X_1+X_2}(t) = f_{X_1}(t) \cdot f_{X_2}(t) = e^{ia_1 t - \sigma_1^2 t^2/2} \cdot e^{ia_2 t - \sigma_2^2 t^2/2}$$
$$= e^{i(a_1+a_2)t - (\sigma_1^2 + \sigma_2^2)t^2/2}.$$

Furthermore, if $P_X = \mathcal{N}(a, \sigma^2)$, then for any $k \in \mathbb{R}$ we have

$$P_{kX} = \mathcal{N}(ka, k^2\sigma^2), \tag{2.17}$$

since

$$f_{kX}(t) = f_X(kt) = e^{iakt - \sigma^2 k^2 t^2/2} = e^{i(ka)t - (k\sigma)^2 t^2/2}.$$

In particular, by combining (2.17) with (2.16) we derive that for any independent $\mathcal{N}(a, \sigma^2)$-distributed variables X_1, X_2, X it is true that

$$P_{k_1 X_1 + k_2 X_2} = \mathcal{N}((k_1 + k_2)a, (k_1^2 + k_2^2)\sigma^2),$$

$$P_{k_3 X + k_4} = \mathcal{N}((k_3 a + k_4, k_3^2 \sigma^2).$$

Therefore, we obtain stability with $k_3 := (k_1^2 + k_2^2)^{1/2}$, $k_4 := (k_1 + k_2 - k_3)a$. Moreover, a normal X is strictly stable iff $a = 0$.

Exercise 2.5. Using the formula for characteristic functions (1.18) prove the strict stability of Cauchy distribution. Namely, for any $\mathcal{C}(a, \sigma)$-distributed independent variables X_1, X_2 it is true that

$$P_{k_1 X_1 + k_2 X_2} = \mathcal{C}((k_1 + k_2)a, (k_1 + k_2)\sigma) = P_{(k_1+k_2)X_1}.$$

In general, one can show that for a stable variable X the parameter k_3 may depend on k_1 and k_2 only in a very special way, namely,

$$k_3 = (k_1^\alpha + k_2^\alpha)^{1/\alpha} \tag{2.18}$$

where $\alpha \in (0, 2]$. In this case X is called α-*stable*. We have just seen that normal variables and distributions are 2-stable. Actually, those are the only existing 2-stable objects. In the sequel, we will be mostly focused on *non*-Gaussian case $0 < \alpha < 2$.

Let us fix an $\alpha \in (0, 2)$, $a \in \mathbb{R}$, and a pair of non-negative numbers c_-, c_+ such that at least one of them is strictly positive. Consider an infinitely divisible variable X with a triplet $(a, 0, \nu)$, where the intensity measure $\nu = \nu(\alpha, c_-, c_+)$ is given by the density $\frac{c(u)}{|u|^{\alpha+1}}$ and

$$c(u) = \begin{cases} c_-, & u < 0, \\ c_+, & u > 0. \end{cases}$$

In other words,

$$\mathbb{E}\, e^{itX} = \exp\left\{ iat + \int (e^{itu} - 1 - itu\,\mathbf{1}_{\{|u|\leq 1\}}) \frac{c(u)du}{|u|^{\alpha+1}} \right\}. \qquad (2.19)$$

Note that the measure ν satisfies the crucial Lévy-Khinchin condition (2.12) iff $\alpha \in (0, 2)$.

We use the notation $\mathcal{S}(\alpha, c_-, c_+, a)$ for the distribution of X. Parameters α, c_-, c_+ reflect the fundamental properties of the distribution while a has no particular meaning because it depends on the arbitrary choice of sets $\mathcal{I} = [-1, 1]$ and $\mathcal{O} = \mathbb{R} \backslash \mathcal{I}$ in the procedure of partial centering.

We will show now that X is stable (in fact, *any* stable non-Gaussian variable has this form of distribution). First, let us describe the distribution of kX with $k > 0$. By using (2.19) we have

$$\mathbb{E}\, e^{it(kX)} = \mathbb{E}\, e^{i(kt)X} = \exp\left\{ iakt + \int (e^{iktu} - 1 - iktu\,\mathbf{1}_{\{|u|\leq 1\}}) \frac{c(u)du}{|u|^{\alpha+1}} \right\}$$

$$= \exp\left\{ iakt + \int (e^{itv} - 1 - itv\,\mathbf{1}_{\{|v|\leq k\}}) \frac{k^\alpha c(v)dv}{|v|^{\alpha+1}} \right\}$$

$$= \exp\left\{ iakt - it(c_+ - c_-)\beta(k, \alpha) + \int (e^{itv} - 1 - itv\,\mathbf{1}_{\{|v|\leq 1\}}) \frac{k^\alpha c(v)dv}{|v|^{\alpha+1}} \right\}$$

where

$$\beta(k, \alpha) = k^\alpha \int_1^k \frac{dv}{v^\alpha} = \begin{cases} \frac{k - k^\alpha}{1 - \alpha}, & \alpha \neq 1; \\ k \ln k, & \alpha = 1. \end{cases}$$

In other words, $P_{kX} = \mathcal{S}(\alpha, k^\alpha c_-, k^\alpha c_+, ka - (c_+ - c_-)\beta(k, \alpha))$. From this calculation we retain, by the way, that the case $\alpha = 1$ is somewhat special.

Next, by addition rule for triplets, we see that for any $k_1, k_2 > 0$ and independent copies X_1, X_2 of X it is true that

$$P_{k_1 X_1 + k_2 X_2} = \mathcal{S}(\alpha, (k_1^\alpha + k_2^\alpha)c_-, (k_1^\alpha + k_2^\alpha)c_+, k_{12})$$

with $k_{12} := (k_1 + k_2)a - (c_+ - c_-)[\beta(k_1, \alpha) + \beta(k_2, \alpha)]$. On the other hand, for any $k_3 > 0$ and any k_4 we have

$$P_{k_3 X + k_4} = \mathcal{S}(\alpha, k_3^\alpha c_-, k_3^\alpha c_+, (k_3 a + k_4 - (c_+ - c_-)\beta(k_3, \alpha))).$$

Hence, by taking k_3 from (2.18) and letting

$$k_4 = k_{12} - k_3 a + (c_+ - c_-)\beta(k_3, \alpha)$$

we obtain

$$P_{k_1 X_1 + k_2 X_2} = P_{k_3 X + k_4}.$$

Therefore, X is stable.

Let us now see in which cases X is *strictly* stable. In other words, we have to check when $k_4 = 0$ holds for any k_1, k_2. By using the definitions of k_4 and k_{12}, the equation $k_4 = 0$ boils down to

$$(k_1 + k_2 - k_3)a = (c_+ - c_-)\big(\beta(k_1, \alpha) + \beta(k_2, \alpha) - \beta(k_3, \alpha)\big).$$

If $\alpha \neq 1$, then, using the definitions of $\beta(\cdot, \alpha)$ and k_3, we additionally have

$$\beta(k_1, \alpha) + \beta(k_2, \alpha) - \beta(k_3, \alpha) = \frac{k_1 - k_1^\alpha}{1-\alpha} + \frac{k_2 - k_2^\alpha}{1-\alpha} - \frac{k_3 - k_3^\alpha}{1-\alpha} = \frac{k_1 + k_2 - k_3}{1-\alpha},$$

and the strict stability is achieved whenever $a = \frac{c_+ - c_-}{1-\alpha}$. Notice that for any fixed $\alpha \neq 1$, c_-, c_+ there exists a unique a such that the distribution $S(\alpha, c_-, c_+, a)$ is strictly stable.

On the other hand, if $\alpha = 1$, then $k_1 + k_2 - k_3 = 0$, hence the strict stability holds whenever $c_- = c_+$.

In the non-symmetric case $\alpha = 1, c_- \neq c_+$ one can not render X strictly stable by an appropriate shift.

Back to terminology, if $a = 0$, $c_- = c_+$, then X is called *symmetric stable*; if $c_- = 0$ or $c_+ = 0$, then X and its distribution are called *totally skewed stable* or spectrally positive (resp. spectrally negative) stable.

Now we show that characteristic function (2.19) admits more explicit representation.

If $\alpha \in (0, 1)$, then

$$\int_0^1 u^{-\alpha} = \frac{1}{1-\alpha} < \infty,$$

hence we have

$$\mathbb{E}\, e^{itX} = \exp\left\{ i\tilde{a}t + \int (e^{itu} - 1)\,\frac{c(u)\,du}{|u|^{\alpha+1}} \right\}, \tag{2.20}$$

where

$$\tilde{a} := a + \frac{c_+ - c_-}{\alpha - 1}.$$

Notice that X is a shifted variable from the class $\mathcal{P}_{1,0}$. Moreover, X is strictly stable whenever $\tilde{a} = 0$.

By using the well known formulae

$$\int_0^\infty \frac{\sin v}{v^p}\, dv = \frac{\pi}{2\sin(p\pi/2)\Gamma(p)}\,, \qquad 0 < p < 2,$$

and

$$\int_0^\infty \frac{1 - \cos v}{v^{p+1}}\, dv = \frac{\pi}{2\sin(p\pi/2)\Gamma(p+1)}\,, \qquad 0 < p < 2, \tag{2.21}$$

we easily obtain an explicit expression for characteristic function,

$$\mathbb{E}\, e^{itX}$$
$$= \exp\left\{ i\widetilde{a}t - \frac{\pi|t|^\alpha}{2\Gamma(\alpha+1)}\left[\frac{c_+ + c_-}{\sin(\pi\alpha/2)} - \frac{i\,(c_+ - c_-)\mathrm{sgn}(t)}{\cos(\pi\alpha/2)}\right]\right\}, \tag{2.22}$$

where

$$\mathrm{sgn}(t) := \begin{cases} +1, & t > 0 \\ -1, & t < 0 \end{cases}.$$

In particular, for symmetric case $a = 0$, $c_- = c_+$ we obtain a nice expression

$$\mathbb{E}\, e^{itX} = \exp\left\{-K|t|^\alpha\right\} \tag{2.23}$$

with $K := \frac{\pi c_+}{\sin(\pi\alpha/2)\Gamma(\alpha+1)}$.

If $\alpha \in (1,2)$, then

$$\int_1^\infty u^{-\alpha} = \frac{1}{\alpha - 1} < \infty,$$

hence we have

$$\mathbb{E}\, e^{itX} = \exp\left\{ i\widetilde{a}t + \int (e^{itu} - 1 - itu)\,\frac{c(u)du}{|u|^{\alpha+1}}\right\}, \tag{2.24}$$

with the same $\widetilde{a}$ as above.

We see that X is a shifted variable from the class $\mathcal{P}_{2,1}$. Notice that X is strictly stable whenever $\widetilde{a} = 0$, or, equivalently, $\mathbb{E}\, X = 0$. (For the former case $\alpha < 1$, it was not possible to state a strict stability condition in terms of $\mathbb{E}\, X$ because expectation just does not exist there.)

By using (2.21) along with integration by parts formula

$$\int_0^\infty \frac{\sin v - v}{v^{p+1}}\, dv = -\int_0^\infty \frac{1 - \cos v}{p\, v^p}\, dv$$
$$= \frac{\pi}{2\cos(p\pi/2)\Gamma(p+1)}\,, \qquad 1 < p < 2,$$

we arrive, somewhat unexpectedly, to the same formulae (2.22) and (2.23) for characteristic function.

Finally, let $\alpha = 1$, $a = 0$, $c_- = c_+$. Then (2.21) yields

$$\int (e^{itu} - 1 - itu\, \mathbf{1}_{\{|u| \le 1\}}) \frac{c(u)du}{|u|^{\alpha+1}} = -2c_+ \int_0^\infty (1 - \cos(tu)) \frac{du}{u^2}$$
$$= -\pi c_+ |t|, \qquad (2.25)$$

in full agreement with (2.23). Expression (2.25) clearly corresponds to Cauchy distribution $\mathcal{C}(0, \pi c_+)$, cf. (1.18). In particular, the triplet of the standard Cauchy distribution is $a = 0$, $\alpha = 1$, $\nu(du) = \frac{du}{\pi u^2}$.

Exercise 2.6. Find an explicit expression for characteristic function in the non-symmetric case $\alpha = 1, c_+ \ne c_-$.

One can read much more about stable variables, processes and limit theorems in the books by Ibragimov and Linnik [41], Taqqu and Samorodnitsky [89], and Zolotarev [117].

3 Limit Theorems for Sums and Domains of Attraction

Let $X, X_1, X_2, \ldots$ be a sequence of i.i.d. random variables. Let denote $S_n := \sum_{j=1}^n X_j$. We say that X belongs to the *domain of attraction* of a distribution $\mathcal{S}$, if for some sequences $B_n > 0$ and $A_n \in \mathbb{R}$ the convergence in distribution holds,

$$\frac{S_n - A_n}{B_n} \Rightarrow \mathcal{S}, \qquad n \to \infty. \qquad (3.1)$$

Notice that the attracting distribution is defined up to a shift and scaling transformations.

The first classical result shows that any random variable with finite variance belongs to the domain of attraction of a normal distribution.

Theorem 3.1 (Lévy CLT). [2] *Let $X, X_1, X_2, \ldots$ be a sequence of i.i.d. random variables such that $\mathbb{E} X^2 < \infty$. Let $a := \mathbb{E} X$, $\sigma^2 := \mathrm{Var} X > 0$. Define $A_n := na$; $B_n := \sigma\sqrt{n}$. Then the distributions of $\frac{S_n - A_n}{B_n}$ converge to $\mathcal{N}(0, 1)$.*

Notice, however, that there exist some variables with infinite variance that also belong to the domain of attraction of normal distributions.

---

[2]CLT – central limit theorem. There is no common agreement about the meaning of the word "central" in this expression.

The domain of attraction of any α-stable distribution with $\alpha < 2$ also is non-empty but it is rather small. In order to belong to such domain a random variable must have sufficiently regularly varying tail probabilities $\mathbb{P}\{X > x\}$ and $\mathbb{P}\{X < -x\}$, as $x \to +\infty$. Moreover, the decay of the "left" and "right" tails must be of the same order.

Before we state a precise result, we need an important auxiliary notion. A function $L : \mathbb{R}_+ \mapsto \mathbb{R}_+$ is called *slowly varying* at infinity if for any $c > 0$ it is true that

$$\lim_{x \to +\infty} \frac{L(cx)}{L(x)} = 1.$$

This essentially means that $L(\cdot)$ varies slower than any power function. For example, the functions $L(x) := [\ln(1 + x)]^\beta$, with $\beta > 0$, and the functions $L(x) := \exp\{C[\ln(1 + x)]^\beta\}$, with $C > 0$, $\beta \in (0, 1)$, are slowly varying at infinity.

Now we are able to describe the domain of attraction of a stable non-Gaussian distribution.

Theorem 3.2. *Let* $0 < \alpha < 2$. *A random variable X belongs to the domain of attraction of a stable distribution* $\mathcal{S}(\alpha, c_-, c_+, a)$, *iff*

$$\mathbb{P}\{X > x\} \sim c_+\, L(x)\, x^{-\alpha}, \qquad x \to +\infty, \qquad (3.2)$$

$$\mathbb{P}\{X < -x\} \sim c_-\, L(x)\, x^{-\alpha}, \qquad x \to +\infty, \qquad (3.3)$$

where $L(x)$ is a slowly varying function.

If X belongs to the domain of attraction of $\mathcal{S}(\alpha, c_-, c_+, a)$, then the norming sequence B_n can be chosen in "arbitrary" way so that

$$\lim_{n \to \infty} n \cdot \mathbb{P}\{|X| > B_n\} = 1.$$

For example, one can take

$$B_n := \sup\left\{x > 0 : \mathbb{P}\{|X| > B_n\} \geq \frac{1}{n}\right\}.$$

According to (3.2) and (3.3) B_n can be written as

$$B_n = [(c_- + c_+)n]^{1/\alpha} L^-(n),$$

where $L^-(\cdot)$ is some other slowly varying function[3].

---

[3]For a curious reader: if $G(\cdot)$ is a slowly varying function, there exists a *dual* slowly varying function $G^*(\cdot)$ such that

$$\lim_{x \to \infty} G^*(x)G(xG^*(x)) = 1,$$

see [91], Theorem 1.5. Let $G(x) := L(x)^{-1/\alpha}$. Then we may set $L^-(n) := G^*(n^{1/\alpha})$.

As for the centering sequence A_n, it depends on α. If $\alpha > 1$, a usual centering by expectation $A_n = n\,\mathbb{E}\,X$ works. For $\alpha < 1$ the centering is not needed, we may let $A_n = 0$. For $\alpha = 1$ the situation, as usual, is more complicated because the centering sequence is not linear in n. One may let

$$A_n = n\,B_n\,\mathbb{E}\left(\frac{X}{X^2 + B_n^2}\right)$$

in this case.

Let us mention one special case. If $L(x) = L = const$, then we say that X belongs to the *domain of normal attraction* of the distribution $\mathcal{S}(\alpha, c_-, c_+, a)$. Here "normal" has nothing to do with normal distributions.

In case of normal attraction the norming factor has a particularly simple power form $B_n = [(c_- + c_+)Ln]^{1/\alpha}$.

Notice that a stable variable X belongs to its own domain of attraction because in this case S_n is equidistributed with $n^{1/\alpha}X + A_n$ with appropriate $A_n \in \mathbb{R}$. Hence X satisfies (3.1) with $B_n = n^{1/\alpha}$.

Domains of attraction of stable distributions are considered in detail in [41], Chapter 2.6.

Finally notice that the class of distributions having non-empty attraction domains exactly *coincides* with the class of stable distributions. This is how the stable distributions were discovered. This is also an explanation of their importance.

4   Random Vectors

4.1   *Definition*

Similarly to random variables, one can define random objects of a more general nature. If $(\mathcal{R}, \mathcal{W})$ is a measurable space, then any measurable mapping $X : (\Omega, \mathcal{A}) \mapsto (\mathcal{R}, \mathcal{W})$ is called an $\mathcal{R}$-*valued random element*.

The distribution P_X, a measure on $(\mathcal{R}, \mathcal{W})$, and the sigma-field $\mathcal{A}_X$ related to X are defined exactly as before. The former definition of independence of random variables remains meaningful for random elements. Moreover, independence is well defined even for random elements taking values in different spaces.

In particular case of Euclidean space $(\mathcal{R}, \mathcal{W}) = (\mathbb{R}^n, \mathcal{B}^n)$, a random element is called *random vector*.

Exercise 4.1. A mapping $\omega \mapsto X(\omega) = (X_1(\omega), \ldots, X_n(\omega)) \in \mathbb{R}^n$ is a random vector iff all its components $X_1, \ldots, X_n$ are random variables.

For any random vector $X = (X_j) \in \mathbb{R}^n$ one understands the *expectation* component-wise, i.e. $\mathbb{E}\,X := (\mathbb{E}\,X_j) \in \mathbb{R}^n$, assuming that every component has a finite expectation. If $(\cdot, \cdot)$ denotes the scalar product in $\mathbb{R}^n$, i.e.

$$(u, v) := \sum_{j=1}^{n} u_j v_j,$$

then for any $v \in \mathbb{R}^n$ we have

$$\mathbb{E}\,(X, v) = \sum_{j=1}^{n} \mathbb{E}\,X_j v_j = (\mathbb{E}\,X, v). \tag{4.1}$$

As in one-dimensional case, we have the summation rule for expectations

$$\mathbb{E}\,(X + Y) = \mathbb{E}\,X + \mathbb{E}\,Y. \tag{4.2}$$

Furthermore, if $L : R^n \mapsto \mathbb{R}^m$ is a linear operator, then

$$\mathbb{E}\,(LX) = L(\mathbb{E}\,X),$$

because by (4.1) for any $v \in \mathbb{R}^m$ we have

$$\mathbb{E}\,(LX, v) = \mathbb{E}\,(X, L^*v) = (\mathbb{E}\,X, L^*v) = (L(\mathbb{E}\,X), v). \tag{4.3}$$

Recall that $L^* : \mathbb{R}^m \to \mathbb{R}^n$ denotes the operator *dual* to L, i.e. such that for all $u \in \mathbb{R}^n, v \in \mathbb{R}^m$ it is true that

$$(Lu, v) = (u, L^*v).$$

By taking the j-th coordinate vector as v we obtain

$$[\mathbb{E}\,(LX)]_j = \mathbb{E}\,[(LX)_j] = [L(\mathbb{E}\,X)]_j,$$

as required in (4.3).

The *covariance operator* of X denoted by $K_X : \mathbb{R}^n \mapsto \mathbb{R}^n$ is defined by the matrix

$$K_X = (\text{cov}(X_i, X_j))_{1 \le i,j \le n}, \tag{4.4}$$

assuming that every component has a finite variance, hence all covariances are well defined. For any $u, v \in \mathbb{R}^n$ we have

$$\text{cov}((X, u), (X, v)) = \text{cov}\left(\sum_{i=1}^{n} X_i u_i, \sum_{j=1}^{n} X_j v_j, \right)$$

$$= \sum_{i,j=1}^{n} \text{cov}(X_i, X_j) u_i v_j$$

$$= \sum_{i=1}^{n} u_i \sum_{j=1}^{n} \text{cov}(X_i, X_j) v_j$$

$$= \sum_{i=1}^{n} u_i (K_X v)_i = (u, K_X v).$$

Furthermore, any covariance operator is necessarily symmetric and non-negative definite, which means

$$(u, K_X v) = (v, K_X u), \qquad \forall u, v \in \mathbb{R}^n, \tag{4.5}$$

and

$$\sum_{i,j=1}^{n} (K_X)_{ij} c_i \overline{c_j} \geq 0, \qquad \forall c_i \in \mathbb{C}, \tag{4.6}$$

respectively. Indeed, (4.5) follows from the obvious symmetry of the matrix (4.4), while (4.6) follows from

$$\sum_{i,j=1}^{n} (K_X)_{ij} c_i \overline{c_j} = \sum_{i,j=1}^{n} \mathrm{cov}(X_i, X_j) c_i \overline{c_j}$$

$$= \mathrm{cov}\Big(\sum_{i=1}^{n} c_i X_i, \sum_{j=1}^{n} c_j X_j \Big)$$

$$= \mathbb{E} \Big| \sum_{i=1}^{n} c_i (X_i - \mathbb{E}\, X_i) \Big|^2 \geq 0.$$

It is well known that any symmetric non-negatively defined operator admits a *diagonalization*: there exists an orthonormal base (e_j) such that K_X has a diagonal form

$$K_X e_j = \lambda_j\, e_j \qquad \text{with some } \lambda_j \geq 0. \tag{4.7}$$

As in one-dimensional case, we have the summation rule for covariances of *independent* vectors

$$K_{X+Y} = K_X + K_Y. \tag{4.8}$$

Furthermore, if $L : R^n \mapsto \mathbb{R}^m$ is a linear operator, then

$$K_{LX} = LK_X L^*, \tag{4.9}$$

because for any $u, v \in \mathbb{R}^m$ we have

$$\mathrm{cov}((LX, u), (LX, v)) = \mathrm{cov}((X, L^*u), (X, L^*v)) = (L^*u, K_X L^*v)$$
$$= (u, LK_X L^*v).$$

By taking coordinate functionals as u, v, we obtain

$$(K_{LX})_{ij} = \mathrm{cov}((LX)_i, (LX)_j) = (LK_X L^*)_{ij},$$

as required in (4.9).

Characteristic function $f_X : \mathbb{R}^n \mapsto \mathbb{C}$ of a vector X is defined by
$$f_X(t) := \mathbb{E}\, e^{i(t,X)}, \qquad t \in \mathbb{R}^n.$$

It has the same basic properties as the univariate characteristic functions. In particular, if the vectors X and Y are independent, then $f_{X+Y}(t) = f_X(t)f_Y(t)$. Notice also the behavior of f_X by linear transformations: for any $a \in \mathbb{R}^m$ and any linear operator $L : \mathbb{R}^n \mapsto \mathbb{R}^m$ it is true that for any $t \in \mathbb{R}^m$
$$f_{LX+a}(t) = e^{i(t,a)}\mathbb{E}\, e^{i(t,LX)} = e^{i(t,a)}\mathbb{E}\, e^{i(L^*t,X)} = e^{i(t,a)} f_X(L^*t).$$

Similarly to the univariate case, there is an inversion formula for reconstructing of the distribution through its characteristic function. Therefore, characteristic function determines the distribution uniquely.

There is a useful criterion for independence of components of a random vector in terms of its characteristic function. Let an n-dimensional random vector X be obtained by concatenation of m random vectors $Y^1, \dots, Y^m$ as follows. Let
$$0 = n_0 < n_1 < \cdots < n_m = n.$$
For $1 \le q \le m$ take an $(n_q - n_{q-1})$-dimensional random vector Y^q and let
$$X_j := Y^q_{j-n_{q-1}}, \qquad 1 \le q \le n,\ n_{q-1} < j \le n_q.$$
To state is simply, we write down all components of Y^1, then all components of Y^2, etc.

Similarly, for any deterministic vector $t \in \mathbb{R}^n$ let define $t^q \in \mathbb{R}^{n_q - n_{q-1}}$ by
$$t^q_l := t_{l-n_{q-1}}, \qquad 1 \le q \le n,\ 1 \le l \le n_q - n_{q-1}.$$
Then we have
$$f_X(t) = \mathbb{E}\, \exp\{i(t,X)\} = \mathbb{E}\, \exp\Big\{i\sum_{j=1}^{n} t_j X_j\Big\}$$
$$= \mathbb{E}\, \exp\Big\{i\sum_{q=1}^{m}\sum_{l=1}^{n_q-n_{q-1}} t^q_l Y^q_l\Big\}$$
$$= \mathbb{E}\left(\prod_{q=1}^{m} \exp\left\{i(t^q, Y^q)\right\}\right).$$

If the vectors $Y^1, \dots, Y^m$ are independent, then the product falls apart, and we obtain
$$f_X(t) = \prod_{q=1}^{m} \mathbb{E}\, \exp\{i(t^q, Y^q)\} = \prod_{q=1}^{m} f_{Y_q}(t^q). \qquad (4.10)$$

Furthermore, it is not difficult to derive (from the fact that a characteristic function determines its distribution uniquely) that converse is also true: if (4.10) holds, then the vectors $Y^1, \ldots, Y^m$ are independent.

4.2   Convergence of random vectors

A.s. convergence, convergence in the mean and convergence in probability of random vectors are defined exactly as for random variables. Convergence in distribution is now more convenient to define in the way suggested by Theorem 1.21:

Theorem 4.2. *Let* (X_k) *and* X *be some* $\mathbb{R}^n$*-valued random vectors. Then the following assertions are equivalent.*

a) For any closed set $B \subset \mathbb{R}^n$ *it is true that*

$$\limsup_{k \to \infty} \mathbb{P}\{X_k \in B\} \le \mathbb{P}\{X \in B\}.$$

b) For any open set $V \subset \mathbb{R}^n$ *it is true that*

$$\liminf_{k \to \infty} \mathbb{P}\{X_k \in V\} \ge \mathbb{P}\{X \in V\}.$$

c) For any measurable set $A \subset \mathbb{R}^n$, *satisfying* regularity condition *with respect to* X, *i.e.* $\mathbb{P}(X \in \partial A) = 0$, *it is true that*

$$\lim_{k \to \infty} \mathbb{P}\{X_k \in A\} = \mathbb{P}\{X \in A\}.$$

d) For any bounded continuous function $f : \mathbb{R}^n \mapsto \mathbb{R}$ *it is true that*

$$\lim_{k \to \infty} \mathbb{E}\, f(X_k) = \mathbb{E}\, f(X).$$

If any of properties a) – d) holds, we say that the sequence X_k *converges in distribution* or *converges in law* or *converges weakly* to X and write $X_k \Rightarrow X$.   The relations between four types of convergence shown on Figure 1.1 hold true for random vectors.

Exercise 4.3. Let $X_k \Rightarrow X$ in $\mathbb{R}^n$ and let $g : \mathbb{R}^n \to \mathbb{R}^m$ be a continuous function. Then $g(X_k) \Rightarrow g(X)$. Hint: use definition d) from Theorem 4.2.

Exercise 4.4. Let (X_k) and (Y_k) be two sequences of random vectors on a common probability space such that $X_k \Rightarrow X$ and $Y_k \xrightarrow{\mathbb{P}} 0$. Prove that $X_k + Y_k \Rightarrow X$.

The univariate criterion of weak convergence in terms of characteristic functions, see Theorem 4.5, remains valid for random vectors: we have

Theorem 4.5. *Let (X_k) be a sequence of random vectors in $\mathbb{R}^n$ such that*

$$\lim_{k \to \infty} f_{X_k}(t) \to f(t), \qquad \forall t \in \mathbb{R}^n,$$

while the limit function $f(\cdot)$ is continuous at zero. Then there exists a random vector X such that

$$X_k \Rightarrow X,$$

and we have $f_X(\cdot) = f(\cdot)$.

Exercise 4.6. Let (X_k) be a sequence of $\mathbb{R}^n$-valued random vectors and X an $\mathbb{R}^n$-valued random vector. Let X_j^k denote the j-th component of X_k. Assume that for each k the components $(X_j^k)_{j=1}^n$ are independent, and that the component-wise weak convergence holds, i.e. for each $j = 1, 2, \ldots, n$, it is true that

$$X_j^k \Rightarrow X_j, \qquad \text{as } k \to \infty.$$

Then $X_k \Rightarrow X$.

Exercise 4.7. Let (X_k) and X be $\mathbb{R}^n$-valued random vectors such that $X_k \Rightarrow X$. Let $A : \mathbb{R}^n \mapsto \mathbb{R}^m$ be a linear operator. Prove that $AX_k \Rightarrow AX$.

In particular, $X_k \Rightarrow X$ iff the weak convergence of differences holds, i.e. $\Delta_k \Rightarrow \Delta$, where

$$\Delta_k := \begin{cases} X_j^k, & j = 1, \\ X_j^k - X_{j-1}^k, & j = 2, \ldots, n, \end{cases} \qquad \Delta := \begin{cases} X_j, & j = 1, \\ X_j - X_{j-1}, & j = 2, \ldots, n. \end{cases}$$

There is a nice trick that reduces multivariate weak convergence to the univariate one.

Theorem 4.8 (Cramér–Wold criterion). *Let (X_k) be a sequence of random vectors in $\mathbb{R}^n$. Then*

$$X_k \Rightarrow X,$$

iff for any $t \in \mathbb{R}^n$ it is true that $(t, X_k) \Rightarrow (t, X)$.

Proof. If $X_k \Rightarrow X$, then by Theorem 4.2 d) we have $f_{X_k}(v) \to f_X(v)$ for any $v \in \mathbb{R}^n$. By letting $v := ut$ with arbitrary $u \in \mathbb{R}$, we obtain

$$f_{(t,X_k)}(u) = \mathbb{E}\, e^{iu(t,X_k)} = f_{X_k}(ut) \to f_X(ut) = \mathbb{E}\, e^{i(ut,X)} = f_{(t,X)}(u).$$

By Theorem 1.22 we obtain $(t, X_k) \Rightarrow (t, X)$.

Conversely, let $(t, X_k) \Rightarrow (t, X)$ for any $t \in \mathbb{R}^n$. Then for any $u \in \mathbb{R}$ we have $f_{(t,X_k)}(u) \to f_{(t,X)}(u)$. By letting $u := 1$ we have

$$f_{X_k}(t) = \mathbb{E}\, e^{i(t,X_k)} = f_{(t,X_k)}(1) \to f_{(t,X)}(1) = \mathbb{E}\, e^{i(t,X)} = f_X(t).$$

Theorem 4.5 yields $X_k \Rightarrow X$. □

4.3 *Gaussian vectors*

A random vector $Y \in \mathbb{R}^n$ and its distribution are called *Gaussian*, if the scalar product (Y, v) is a normal random variable for each $v \in \mathbb{R}^n$.

One can approach the notion of Gaussian vector more constructively. A random vector $X = (X_j)_{j=1}^n \in \mathbb{R}^n$ and its distribution are called *standard Gaussian*, if the components of X independent and have a standard normal distribution. The distribution of X has a density

$$p_X(x) = \frac{1}{(2\pi)^{n/2}} \exp\left\{\frac{-(x,x)}{2}\right\}, \qquad x \in \mathbb{R}^n.$$

Let $a \in \mathbb{R}^n$, and let $L : \mathbb{R}^n \mapsto \mathbb{R}^n$ be a linear mapping. Then $Y := a + LX$ is a Gaussian vector. Indeed,

$$(a + LX, v) = (a, v) + (X, L^*v) = (a, v) + \sum_{j=1}^n (L^*v)_j X_j \qquad (4.11)$$

has a normal distribution due to summation rule (2.16) for the normal variables.

It is easy to show that every Gaussian vector Y in $\mathbb{R}^n$ admits a representation (4.11) with some $a \in \mathbb{R}^n$, $L : \mathbb{R}^n \mapsto \mathbb{R}^n$ and appropriate standard Gaussian vector X. Moreover, by the transformation rule for densities, if the mapping L is invertible (non-degenerated), then Y has a density

$$
\begin{aligned}
p_Y(y) &= \frac{1}{|\det L|}\, p_X\left(L^{-1}(y - a)\right) \\
&= \frac{1}{|\det L|\, (2\pi)^{n/2}} \exp\left\{\frac{-(L^{-1}(y-a), L^{-1}(y-a))}{2}\right\}, \qquad y \in \mathbb{R}^n.
\end{aligned}
$$

However, in the multivariate setting, a definition of Gaussian distribution through a particular form of the density is much less convenient, because in many cases (when the operator L is degenerated, i.e. its image does not coincide with $\mathbb{R}^n$) the density just does not exist.

Note that defining of a Gaussian vector as $a + LX$ is not appropriate for infinite-dimensional generalizations: in most of interesting spaces there is no standard Gaussian vector X.

Let us stress that there exist non-Gaussian random vectors X such that every component X_j of X is a normal random variable. Indeed, let $n = 2$, let X_1 be a standard normal random variable and let

$$X_2 := \begin{cases} X_1, & |X_1| > 1; \\ -X_1, & |X_1| \leq 1. \end{cases}$$

Clearly, $P_{X_2} = P_{X_1} = \mathcal{N}(0,1)$ but $0 < \mathbb{P}(X_1 + X_2 = 0) < 1$, hence $X_1 + X_2$ is a non-Gaussian random variable and $X = (X_1, X_2)$ is a non-Gaussian random vector.

Similarly to the univariate notation $\mathcal{N}(a, \sigma^2)$ introduced earlier, the family of n-dimensional Gaussian distributions also admits a reasonable parametrization. For a Gaussian vector Y write $P_Y = \mathcal{N}(a, K)$ with $a \in \mathbb{R}^n$, $K : \mathbb{R}^n \mapsto \mathbb{R}^n$, if $\mathbb{E}\, Y = a$ and $K_Y = K$. In particular, for a standard Gaussian vector X we have $P_X = \mathcal{N}(0, E_n)$, where $E_n : \mathbb{R}^n \mapsto \mathbb{R}^n$ is the identity operator.

Proposition 4.9.

a) Let Y be a Gaussian vector. Then $P_Y = \mathcal{N}(a, K)$ for some $a \in \mathbb{R}^n$ and some non-negative definite and symmetric operator K.

b) For any $a \in \mathbb{R}^n$ and any non-negative definite and symmetric operator K there exists a vector Y such that $P_Y = \mathcal{N}(a, K)$.

c) The distribution $\mathcal{N}(a, K)$ is unique.

Proof.

a) All components of a Gaussian vector are normal random variables, hence they have finite variances. Therefore, any Gaussian vector has an expectation and a covariance operator, i.e. any Gaussian distribution can be written in the form $\mathcal{N}(a, K)$.

b) Let $a \in \mathbb{R}^n$, and let K be a non-negative definite and symmetric linear operator. Consider a base (e_j) corresponding to the diagonal form of K (see (4.7)) and define $L = K^{1/2}$ by relations $Le_j = \lambda_j^{1/2} e_j$. Then $L^2 = K$. Further, take a standard Gaussian vector X and let $Y = a + LX$. We have already seen that Y is Gaussian. Clearly,

$$\mathbb{E}\, Y = a + \mathbb{E}\,(LX) = a + L(\mathbb{E}\, X) = a + 0 = a$$

and by (4.9)

$$K_Y = K_{LX} = LK_X L^* = LE_n L^* = LL^* = L^2 = K.$$

Hence, by definition $P_Y = \mathcal{N}(a, K)$.

c) Take any $t \in \mathbb{R}^n$. The random variable (Y, t) is normal; it has expectation (a, t) and the variance

$$\mathbb{V}ar(Y, t) = cov((Y, t), (Y, t)) = (t, Kt).$$

Hence, $P_{(Y,t)} = \mathcal{N}((a, t), (t, Kt))$. It follows from the univariate formula for characteristic function (1.17) that

$$\mathbb{E}\, e^{i(Y,t)} = e^{i(a,t) - (t,Kt)/2}. \tag{4.12}$$

Since the characteristic function determines the distribution uniquely, we conclude that a and K determine the distribution uniquely. $\square$

Let us mention few properties of Gaussian vectors that follow straightforwardly from their univariate counterparts.

Consider first summation of independent vectors. If X and Y are independent Gaussian vectors such that $P_X = \mathcal{N}(a_1, K_1)$ and $P_Y = \mathcal{N}(a_2, K_2)$, then $X + Y$ also is a Gaussian vector and $P_{X+Y} = \mathcal{N}(a_1 + a_2, K_1 + K_2)$.

Indeed, for any $v \in \mathbb{R}^n$ we have $(v, X + Y) = (v, X) + (v, Y)$ which is a normal random variable as a sum of independent normal variables. Hence, $X + Y$ is a Gaussian vector. By the general summation rules for expectations (4.2) and covariance operators (4.8) we have equalities $\mathbb{E}\,(X + Y) = a_1 + a_2$, $K_{X+Y} = K_1 + K_2$. It follows that

$$P_{X+Y} = \mathcal{N}(a_1 + a_2, K_1 + K_2).$$

The Gaussian property is also preserved by linear transformations: if $L : \mathbb{R}^n \mapsto \mathbb{R}^m$ is a linear operator, $h \in \mathbb{R}^m$, and $P_X = \mathcal{N}(a, K)$, then the vector $h + LX$ is Gaussian and

$$P_{h+LX} = \mathcal{N}(h + La, LKL^*).$$

Indeed, for any $v \in \mathbb{R}^m$ we have $(v, h + LX) = (v, h) + (L^*v, X)$ which is a normal random variable. Hence, $h + LX$ is a Gaussian vector. By the general transformation rules for expectations (4.3) and covariance operators (4.9) we have $\mathbb{E}\,(h + LX) = h + La$, $K_{h+LX} = LKL^*$. It follows that $P_{h+LX} = \mathcal{N}(h + La, LKL^*)$.

We proceed with a useful criterion for independence of Gaussian vectors. Let an n-dimensional random vector X be obtained by concatenation of m random vectors $Y^1, \ldots, Y^m$ as follows. Let

$$0 = n_0 < n_1 < \cdots < n_m = n.$$

For $1 \le q \le m$ take an $(n_q - n_{q-1})$-dimensional random vector Y^q and let

$$X_j := Y^q_{j-n_{q-1}}, \qquad 1 \le q \le n, \ n_{q-1} < j \le n_q.$$

Proposition 4.10. *Assume that X is a Gaussian vector and that the components of different Y's are uncorrelated, i.e.*

$$cov(Y^{q_1}_{l_1}, Y^{q_2}_{l_2}) = 0, \qquad q_1 \ne q_2, \ l_1 \le n_{q_1} - n_{q_1-1}, \ l_2 \le n_{q_2} - n_{q_2-1}.$$

Then the random vectors $Y^1, \ldots, Y^m$ are independent.

Notice that assuming X to be Gaussian we require slightly more than gaussianity of Y^q for every q.

Proof. We will use the independence criterion in terms of characteristic functions (4.10). As before, for any deterministic vector $t \in \mathbb{R}^n$ we let the vector $t^q \in \mathbb{R}^{n_q-n_{q-1}}$ be defined by

$$t^q_l := t_{l-n_{q-1}}, \qquad 1 \le q \le n, \ 1 \le l \le n_q - n_{q-1}.$$

For example, if $a = \mathbb{E}\, X$, then $a^q = \mathbb{E}\,(Y^q)$.

Furthermore, under assumption of proposition the covariance matrix K_X consists of m diagonally located blocks, where each of these blocks is the covariance matrix K_{Y^q}. In particular, this means that

$$(t, K_X t) = \sum_{q=1}^{m} (t^q, K_{Y^q} t^q), \qquad t \in \mathbb{R}^n.$$

Now by using the formula for Gaussian characteristic function (4.12) we have

$$
\begin{aligned}
f_X(t) &= \exp\left\{ i(a, t) - (t, K_X t)/2 \right\} \\
&= \exp\left\{ i \sum_{q=1}^{m} (a^q, t^q) - \sum_{q=1}^{m} (t^q, K_{Y^q} t^q)/2 \right\} \\
&= \prod_{q=1}^{m} \exp\left\{ i(a^q, t^q) - (t^q, K_{Y^q} t^q)/2 \right\} = \prod_{q=1}^{m} f_{Y^q}(t^q),
\end{aligned}
$$

as required in the independence criterion (4.10). It follows that the vectors $Y^1, \ldots, Y^m$ are independent.   □

Quite often, a particular case of this proposition is applied, dealing with random variables instead of random vectors. In other words, if $m = n$ and $n_q = q$ for $q = 0, 1, \ldots, n$ we obtain the following.

Corollary 4.11. *Let X be a Gaussian vector with pairwise uncorrelated components $X_1, \ldots, X_n$. Then these components are independent random variables.*

4.4 Multivariate CLT

Once multivariate Gaussian distributions are introduced, we are in position
to state and prove a multivariate version of Lévy Central Limit Theorem
(CLT).

Theorem 4.12 (Multivariate Lévy CLT). *Let $X_1, X_2, \ldots$ be a sequence of independent identically distributed n-dimensional random vectors with common finite expectation a and covariance operator K. Let
$Z_m := \frac{\sum_{k=1}^{m} X_k - ma}{\sqrt{m}}$. Then, as $m \to \infty$,*

$$P_{Z_m} \Rightarrow \mathcal{N}(0, K).$$

Proof. Let Y be a Gaussian random vector with distribution $\mathcal{N}(0, K)$.
According to Theorem 4.8, it is sufficient to check that for any $t \in \mathbb{R}^n$
convergence

$$(Z_m, t) \Rightarrow (Y, t) \tag{4.13}$$

holds. Recall that (Y, t) is a normally distributed random variable with
expectation 0 and variance $\sigma_t^2 := (t, Kt)$. Hence, $P_{(Y,t)} = \mathcal{N}(0, \sigma_t^2)$. On the
other hand,

$$(Z_m, t) = \sigma_t \frac{\sum_{k=1}^{m}(X_k, t) - m(a, t)}{\sigma_t \sqrt{m}},$$

and the fraction is a properly normalized sum of i.i.d. random variables
with finite variance σ_t^2. Univariate CLT (cf. Theorem 3.1) implies that
$P_{(Z_m, t)}$ converge to $\mathcal{N}(0, \sigma_t^2)$, and we are done with (4.13).   □

If covariance operator K is invertible, one can rewrite CLT so that
the *standard* Gaussian distribution appears in the limit. Recall that K is
invertible iff in its diagonal form (4.7) we have $\lambda_j > 0$ for all $j \leq n$. Take
the base (e_j) from (4.7) and define operator $K^{-1/2}$ by

$$K^{-1/2}e_j := \lambda_j^{-1/2} e_j, \qquad 1 \leq j \leq n.$$

We claim that

$$P_{K^{-1/2}Z_m} \Rightarrow \mathcal{N}(0, E_n), \qquad \text{as } m \to \infty.$$

Indeed, in this case we have convergence of characteristic functions

$$\mathbb{E} \exp\{i(t, K^{-1/2}Z_m)\} = \mathbb{E} \exp\{i(K^{-1/2}t, Z_m)\}$$
$$\to e^{-(KK^{-1/2}t, K^{-1/2}t)/2} = e^{-(t,t)/2},$$

in agreement with (4.12), where we substitute $a = 0$, $K = E_n$.

4.5   *Stable vectors*

A random vector $Y \in \mathbb{R}^n$ is called α-*stable*, if the scalar product (Y, v) is an α-stable random variable for each $v \in \mathbb{R}^n$.

For example, all Gaussian random vectors are 2-stable.

Exercise 4.13. a) Prove that any random vector $X = (X_j)_{j=1}^n \in \mathbb{R}^n$ having independent α-stable components is α-stable.

b) Prove that if $Y \in \mathbb{R}^n$ is an α-stable random vector and $L : \mathbb{R}^n \mapsto \mathbb{R}^m$ is a linear operator, $h \in \mathbb{R}^m$, then the vector $h + LX$ is an α-stable random vector in $\mathbb{R}^m$.

In Gaussian case by combining statements a) and b) one can obtain *any* Gaussian vector as a linear image of a Gaussian vector with independent components but this is not the case for α-stable random vectors with $\alpha < 2$.

Nontrivial examples of α-stable random vectors are given by the n-tuples of integrals with respect to an independently scattered stable random measure, see Subsection 7.5 in the next chapter.

We refer to the monographs [89] and [63] for a deep treatment of stable non-Gaussian random vectors.

Chapter 2

Random Processes

5 Random Processes: Main Classes

Let T be an arbitrary set called *parameter set*. Its elements may be interpreted as time instants, geographical locations, etc. A family of random variables $X(t), t \in T$, defined on a common probability space and parameterized by elements of T is called a *random process* or *random function*, or *stochastic process*. If the parameter set T belongs to a multi-dimensional Euclidean space, then X is also called a *random field*.

We denote the distribution of the random vector $(X(t_1), ..., X(t_n))$ by $P^X_{t_1,...,t_n}$. Such distributions are called *finite-dimensional distributions* of the process X. They determine all properties of X completely. If for two processes $X(t), t \in T$, and $Y(t), t \in T$, all finite-dimensional distributions coincide, we say that X and Y are equal in distribution.

Let us define some classes of random processes.

A process $X(t), t \in \mathbb{R}$, is called a *stationary process* if a time shift does not affect its properties. In other words, for any $s \in \mathbb{R}$ the process Y_s defined by $Y_s(t) := X(s + t)$ is equal in distribution to X, i.e. for any $t_1, \ldots t_n \in \mathbb{R}$ we have

$$P^X_{t_1,...,t_n} = P^X_{s+t_1,...,s+t_n}.$$

The definition of a stationary process trivially extends to the processes parameterized by $\mathbb{R}^d$ with arbitrary $d > 1$, and, even more generally, by a group.

The differences of the form $X(t) - X(s)$ are called *increments* of a process X. A process $X(t), t \in T$, where $T = \mathbb{R}$ or $T = \mathbb{R}_+$, is called a *process with stationary increments* if a time change does not affect the

47

properties of its increments. In other words, for all $s \in T$ the processes Y_s defined by $Y_s(t) := X(s + t) - X(s)$ are equal in distribution.

Notice that any stationary process has stationary increments but the converse is not true at all, see Example 5.2 below.

Exercise 5.1. Let X be a stationary process. Prove that its integral,

$$Y(t) := \int_0^t X(s)ds$$

is a process with stationary increments.

The definition of a process with stationary increments trivially extends to the processes parameterized by $\mathbb{R}^d$.

A process $X(t), t \in \mathbb{R}_+$, is called a *process with independent increments* if for any $t_0 \leq t_1 \leq \cdots \leq t_n$ the increments $(X(t_j) - X(t_{j-1}))_{1 \leq j \leq n}$ are independent.

Example 5.2. Let $\lambda > 0$. A process $N(t), t \in \mathbb{R}^+$, is called *Poisson process* of intensity λ if it has independent and stationary increments and its values have Poisson distribution: $P_{N(t)} = \mathcal{P}(\lambda t)$ for all $t \in \mathbb{R}_+$.

The trajectories of a Poisson process, i.e. the random functions $t \mapsto N(t)$ start at zero at time zero, i.e. $N(0) = 0$, they are piecewise constant functions taking only integer values (since Poisson distribution is concentrated on non-negative integers), increasing by jumps of size 1. The times between the jumps are random. A typical trajectory is represented on Figure 5.1. The dashed line on the figure corresponds to the expectation $\mathbb{E} N(t) = \lambda t$. A typical trajectory of a *centered* Poisson process $\widetilde{N}(t) := N(t) - \lambda t$ is represented on Figure 5.2. We refer to [50] for extensive theory of Poisson processes.

Let $H > 0$. A process $X(t), t \in T$, where $T = \mathbb{R}$ or $T = \mathbb{R}_+$ is called an *H-self-similar process* if a linear time change leads to a power scaling of X. In other words, for any $c > 0$ the process Y_c defined by $Y_c(t) := \frac{X(ct)}{c^H}$ is equal in distribution to X. See [31] for more examples and properties of this class. The definition of a self-similar process trivially extends to the processes parameterized by $\mathbb{R}^d$ or by $\mathbb{R}_+^d$ with arbitrary $d > 1$.

Assume now that the values of a process X have finite variances at any time, i.e. $\mathbb{V}ar X(t) < \infty$ for all $t \in T$. Then the functions $a_X : T \mapsto \mathbb{R}$ and $K_X : T \times T \mapsto \mathbb{R}$ defined by

$$a_X(t) := \mathbb{E} X(t);$$
$$K_X(s, t) := \text{cov}(X(s), X(t)),$$

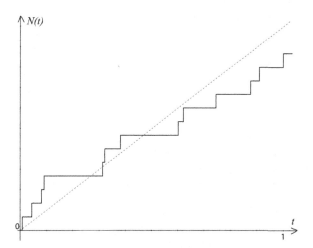

Fig. 5.1 Trajectory of a Poisson process.

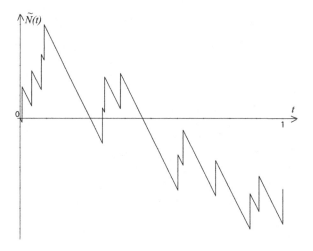

Fig. 5.2 Trajectory of a centered Poisson process.

are called *expectation* and *covariance function* of X, respectively.

For example, by properties of Poisson distribution (1.7), (1.13), and by independence of increments, a Poisson process of intensity λ satisfies

$$a_X(t) = \lambda t,$$

and, for $s \leq t$,

$$K_X(s,t) = \text{cov}(X(s), X(t) - X(s) + X(s))$$
$$= \text{cov}(X(s), X(t) - X(s)) + \text{cov}(X(s), X(s))$$
$$= \text{cov}(X(s) - X(0), X(t) - X(s)) + \mathbb{V}ar(X(s), X(s))$$
$$= 0 + \lambda s = \lambda \min\{s, t\}. \tag{5.1}$$

A process $X(t), t \in T$, is called a *Gaussian process* if all its finite-dimensional distributions $P^X_{t_1,\ldots,t_n}$ are Gaussian.

In particular, every value $X(t)$ of a Gaussian process is a normally distributed random variable. Since they have finite second moment, every Gaussian process has expectation a_X and covariance K_X. Moreover, the finite-dimensional distributions of a Gaussian process X are completely determined by the pair (a_X, K_X). Indeed, for any set $t_1, \ldots, t_n$ in T this pair determines the expectation and the covariance matrix (4.4) of the Gaussian vector $(X(t_1), \ldots, X(t_n))$. Yet we know from Proposition 4.9 that a Gaussian distribution in $\mathbb{R}^n$ is uniquely determined by the expectation and covariance operator.

For general processes, however, a_X and K_X do not determine at all the finite-dimensional distributions of X. For example, a centered Poisson process $X(t) - \lambda t$ with $\lambda = 1$ and a Wiener process described below are completely different but they have the same zero expectation and the same covariance (5.1).

Many examples of Gaussian processes are given in the next section.

A process $X(t), t \in T$, is called α-*stable process* if all its finite-dimensional distributions $P^X_{t_1,\ldots,t_n}$ are α-stable.

In particular, every value $X(t)$ of such process is an α-stable random variable. Since for $\alpha < 2$ the α-stable variables have infinite variances, there is no such simple description for the corresponding α-stable processes as we have for the Gaussian ones. Therefore, the stable processes are mainly described through their integral representations. A typical example of this kind is given below in Subsection 13.5.1 of Chapter 3. Lévy stable process considered in Subsection 9.3 form the most popular but by far not exhaustive subclass of stable processes.

6 Examples of Gaussian Random Processes

Recall that a Gaussian process $X(t), t \in T$, introduced in Section 5, is completely determined by its expectation $a_X(t) = \mathbb{E}\,X(t)$, $t \in T$, and its covariance function $K_X(s,t) = \text{cov}(X(s), X(t))$, $s, t \in T$. We will now consider the most important examples. In all of them expectation vanishes, thus every time we define a process by indicating its covariance function.

6.1 *Wiener process*

Wiener process, or, equivalently, *Brownian motion*, $W(t), 0 \le t < \infty$, is a Gaussian process satisfying assumptions

$$\mathbb{E}\,W(t) = 0, \qquad K_W(s,t) := \mathbb{E}\,W(s)W(t) = \min\{s,t\}.$$

A typical trajectory of a Wiener process is represented on Figure 6.1. Notice the highly irregular behavior of trajectories. Actually they are nowhere differentiable with probability one.

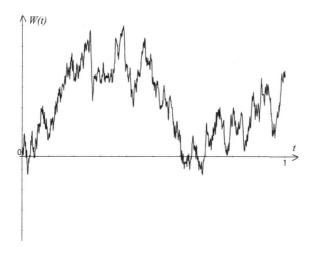

Fig. 6.1   Trajectory of a Wiener process.

Let us discuss the basic properties of a Wiener process.

- *Alternative definition*: a Gaussian process $W(t), 0 \le t < \infty$, is a Wiener process if and only if it satisfies

$$\mathbb{E}\,W(t) = 0, \qquad t \ge 0; \qquad (6.1)$$

$$W(0) = 0; \qquad (6.2)$$

$$\mathbb{E}\,(W(s) - W(t))^2 = |s - t|, \qquad s, t \ge 0. \qquad (6.3)$$

Assume first that W is a Wiener process. Then (6.1) holds by definition. Since $\mathbb{E}\,W(0)^2 = K_W(0,0) = 0$, we obtain (6.2). Finally,

$$
\begin{aligned}
\mathbb{E}\,(W(s) - W(t))^2 &= \mathbb{E}\,W(s)^2 + \mathbb{E}\,W(t)^2 - 2\mathbb{E}\,(W(s)W(t)) \\
&= s + t - 2\min\{s,t\} \\
&= \min\{s,t\} + \max\{s,t\} - 2\min\{s,t\} \\
&= \max\{s,t\} - \min\{s,t\} = |s - t|,
\end{aligned}
$$

as required in (6.3). Conversely, let equalities (6.1)-(6.3) be true. Then by plugging $t = 0$ in (6.3) and using (6.2) we obtain $\mathbb{E}\,W(s)^2 = s$. Similarly, $\mathbb{E}\,W(t)^2 = t$. Finally,

$$
\begin{aligned}
K_W(s,t) &= \mathbb{E}\,(W(s)W(t)) \\
&= \frac{1}{2}\mathbb{E}\,\left(W(s)^2 + W(t)^2 - (W(s) - W(t))^2\right) \\
&= \frac{1}{2}\,(s + t - |s - t|) = \min\{s,t\},
\end{aligned}
$$

as required in the definition of Wiener process.

- $1/2$-*self-similarity*: for any $c > 0$ consider $Y(t) := \frac{W(ct)}{\sqrt{c}}$. Then the process Y is also a Wiener process.

  Indeed, it is clear that Y inherits from W the property of being a Gaussian process. Moreover, we have $\mathbb{E}\,Y(t) = \frac{\mathbb{E}\,W(ct)}{\sqrt{c}} = 0$ and

$$
\begin{aligned}
K_Y(s,t) := \mathbb{E}\,Y(s)Y(t) &= \frac{\mathbb{E}\,(W(cs)W(ct))}{\sqrt{c}\sqrt{c}} \\
&= \frac{\min\{cs, ct\}}{c} = \frac{c\min\{s,t\}}{c} = \min\{s,t\},
\end{aligned}
$$

  as required.

- *Stationary increments*: for any $t_0 \geq 0$ the process

$$
Y(t) := W(t_0 + t) - W(t_0)
$$

is also a Wiener process; Again, it is clear that Y inherits from W the property of being a Gaussian process. Moreover, all features of alternative definition (6.1)-(6.3) are present. Indeed, we have

$$
\begin{aligned}
\mathbb{E}\,(Y(s) - Y(t))^2 &= \mathbb{E}\,(W(t_0 + s) - W(t_0 + t))^2 \\
&= |(t_0 + s) - (t_0 + t)| = |s - t|,
\end{aligned}
$$

as required in (6.3).

- *Independent increments*: let

$$0 = t_0 < t_1 < \cdots < t_n.$$

Then the increments $\big(W(t_{j+1}) - W(t_j)\big)_{0 \leq j < n}$ are independent random variables.

Indeed, since the increments form a Gaussian vector, by Corollary 4.11 it is enough to show that they are uncorrelated. In fact, for any $0 \leq i < j < n$ we have

$$
\begin{aligned}
&\operatorname{cov}\big(W(t_{i+1}) - W(t_i), W(t_{j+1}) - W(t_j)\big) \\
&= \mathbb{E}\big(W(t_{i+1}) - W(t_i)\big)\big(W(t_{j+1}) - W(t_j)\big) \\
&= \mathbb{E}\big(W(t_{i+1})W(t_{j+1})\big) - \mathbb{E}\big(W(t_{i+1})W(t_j)\big) \\
&\quad - \mathbb{E}\big(W(t_i)W(t_{j+1})\big) + \mathbb{E}\big(W(t_i)W(t_j)\big) \\
&= \min\{t_{i+1}, t_{j+1}\} - \min\{t_{i+1}, t_j\} - \min\{t_i, t_{j+1}\} + \min\{t_i, t_j\} \\
&= t_{i+1} - t_{i+1} - t_i + t_i = 0,
\end{aligned}
$$

and independence follows.

- *Time inversion property*: the process

$$
Z(t) := \begin{cases} t\, W(\tfrac{1}{t}), & t > 0, \\ 0, & t = 0, \end{cases}
$$

is also a Wiener process. Indeed, it is clear that Z is a Gaussian process and we have $\mathbb{E}\, Z(t) = t\, \mathbb{E}\, W(\tfrac{1}{t}) = 0$. Moreover,

$$
\begin{aligned}
K_Z(s,t) &:= \mathbb{E}\big(Z(s)Z(t)\big) = st\, \mathbb{E}\big(W(\tfrac{1}{s})W(\tfrac{1}{t})\big) \\
&= st \min\{\tfrac{1}{s}, \tfrac{1}{t}\} = st\, \frac{1}{\max\{s,t\}} = \min\{s,t\},
\end{aligned}
$$

as required.

- *Markov property*. Let us take some instant $t_0 \geq 0$ as "the present time". Then we can represent "the future" of Wiener process $(W(t_0 + t))_{t \geq 0}$ as

$$W(t_0 + t) = W(t_0) + \big(W(t_0 + t) - W(t_0)\big),$$

where the fist term, a random constant, depends on the present location of W, while the second term is independent from the past $(W(s))_{0 \leq s \leq t_0}$ due to the independence of increments.

The combination of such wonderful properties is, essentially unique, as the following two propositions show.

Proposition 6.1. *Let $X(t), t \geq 0$, be a square-mean continuous Gaussian process with stationary and independent increments, satisfying $X(0) = 0$. Then there exist $c \in \mathbb{R}$, $\sigma \geq 0$ and a Wiener process W such that*

$$X(t) = ct + \sigma W(t), \qquad t \geq 0.$$

Proof. Let $a(t) := \mathbb{E} X(t)$. Then $a(\cdot)$ is a continuous additive function, because by stationarity of increments

$$a(s) + a(t) = \mathbb{E} X(s) + \mathbb{E} X(t) = \mathbb{E} X(s) + \mathbb{E} (X(s+t) - X(s))$$
$$= \mathbb{E} X(s+t) = a(t+s).$$

It is well known that continuous additive functions are linear, thus there exists $c \in \mathbb{R}$ such that $a(t) = ct$. Similarly, let $v(t) := \mathbb{V}arX(t)$. Then $v(\cdot)$ is a continuous additive function, because by stationarity and independence of increments for any $s, t \geq 0$

$$v(s) + v(t) = \mathbb{V}arX(s) + \mathbb{V}arX(t)$$
$$= \mathbb{V}ar(X(s) - X(0)) + \mathbb{V}ar(X(s+t) - X(s))$$
$$= \mathbb{V}arX(s+t) = v(t+s).$$

Therefore, there exists $\sigma \geq 0$ such that $v(t) = \sigma^2 t$. If $\sigma = 0$, the variance vanishes and we trivially have $X(t) = \mathbb{E} X(t) = ct$. Let now $\sigma > 0$. Set

$$W(t) := \sigma^{-1}(X(t) - ct).$$

We check that W satisfies (6.1)-(6.3), hence is a Wiener process. The validity of (6.1) and (6.2) is trivial. Finally, we have

$$\mathbb{E} (W(s) - W(t))^2 = \mathbb{V}ar(W(s) - W(t))$$
$$= \sigma^{-2} \mathbb{V}ar(X(s) - X(t)) = \sigma^{-2} \mathbb{V}ar(X(|s - t|)$$
$$= \sigma^{-2} v(|s - t|) = |s - t|,$$

as required in (6.3). $\qquad\qquad\qquad\qquad\qquad\qquad\qquad\qquad\qquad\square$

Proposition 6.2. *Let $X(t), t \geq 0$, be an H-self-similar Gaussian process with independent increments. Then there exist $c \in \mathbb{R}$, $\sigma \geq 0$ and a Wiener process W such that*

$$X(t) = c t^H + \sigma W(t^{2H}), \qquad t \geq 0. \qquad\qquad (6.4)$$

Proof. By self-similarity, we know that for all $k, t \geq 0$ the variables $X(kt)$ and $k^H X(t)$ are equidistributed. By letting $t = 1$ and comparing first the expectations, we have

$$\mathbb{E} X(k) = k^H \mathbb{E} X(1) := c k^H, \qquad k \geq 0. \qquad\qquad (6.5)$$

Similarly, for the variances we have

$$\mathbb{V}arX(k) = k^{2H}\mathbb{V}arX(1) := \sigma^2 k^{2H}, \qquad k \geq 0. \tag{6.6}$$

Notice that for $k = 0$ we have $\mathbb{E}\,X(k) = \mathbb{V}arX(k) = 0$, hence $X(0) = 0$. Now we compute the covariance by using independence of increments. For any $s \leq t$ we have

$$\begin{aligned}
\mathrm{cov}(X(s), X(t)) &= \mathrm{cov}(X(s), X(s) + (X(t) - X(s)) \\
&= \mathrm{cov}(X(s), X(s)) + \mathrm{cov}(X(s) - X(0), X(t) - X(s)) \\
&= \mathbb{V}arX(s) = \sigma^2 s^{2H} = \sigma^2 \min\{s, t\}^{2H}.
\end{aligned}$$

In the trivial case $\sigma = 0$ we just have $\mathbb{V}arX(\cdot) \equiv 0$, hence it is true that $X(t) = \mathbb{E}\,X(t) = ct^H$.

Assume now that $\sigma > 0$. Let

$$W(u) := \sigma^{-1}\left(X(u^{1/2H}) - cu^{1/2}\right).$$

Then we have (6.4), as required. It remains to show that W is a Wiener process. Indeed, the previous calculations yield

$$\mathbb{E}\,W(u) := \sigma^{-1}\left(\mathbb{E}\,X(u^{1/2H} - cu^{1/2}\right) = \sigma^{-1}\left(cu^{1/2} - cu^{1/2}\right) = 0$$

and

$$\mathrm{cov}(W(u), W(v)) := \sigma^{-2}\mathrm{cov}\left(X(u^{1/2H}), X(v^{1/2H})\right) = \min\{u, v\}.$$

which corresponds exactly to Wiener process. $\qquad\square$

The great importance of Wiener process is explained by the basic role it plays in the limit theorems for random processes (invariance principle), see below, as well as in stochastic calculus.

6.2 *Brownian bridge*

Brownian bridge is a Gaussian process $\overset{o}{W}(t), 0 \leq t \leq 1$, satisfying assumptions

$$\mathbb{E}\,\overset{o}{W}(t) = 0, \qquad K_{\overset{o}{W}}(s, t) := \mathbb{E}\,\overset{o}{W}(s)\overset{o}{W}(t) = \min\{s, t\} - st. \tag{6.7}$$

Notice that $K_{\overset{o}{W}}(0, 0) = K_{\overset{o}{W}}(1, 1) = 0$ yields equalities $\overset{o}{W}(0) = \overset{o}{W}(1) = 0$, which somehow explains "bridge" in the process name.

A typical trajectory of a Brownian bridge is represented on Figure 6.2. The irregular local behavior of trajectories is the same as that of Wiener process.

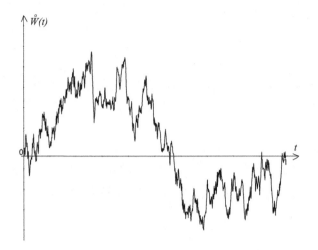

Fig. 6.2   Trajectory of a Brownian bridge.

Next, notice a symmetry in time: if $\overset{o}{W}$ is a Brownian bridge, then the process $Y(t) := \overset{o}{W}(1-t)$ also is a Brownian bridge. Indeed,

$$K_Y(s,t) = K_{\overset{o}{W}}(1-s,1-t) = \min\{1-s,1-t\} - (1-s)(1-t)$$
$$= s + t - \max\{s,t\} - st = \min\{s,t\} - st.$$

Brownian bridge is tightly related to the Wiener process. Namely, let W be a Wiener process. Let

$$\overset{o}{W}(t) := W(t) - tW(1), \qquad 0 \le t \le 1. \tag{6.8}$$

Clearly, $\mathbb{E}\,\overset{o}{W}(t) = 0$ and

$$\mathbb{E}\,\overset{o}{W}(s)\overset{o}{W}(t) = \mathbb{E}\left((W(s) - sW(1))(W(t) - tW(1))\right)$$
$$= \min\{s,t\} - t\min\{s,1\} - s\min\{1,t\} + st\min\{1,1\}$$
$$= \min\{s,t\} - ts - st + st = \min\{s,t\} - st,$$

as required in the definition of $\overset{o}{W}$. Using $W(0) = 0$, we may rewrite (6.8) as $\overset{o}{W}(t) = W(t) - [tW(1) + (1-t)W(0)]$, thus interpreting Brownian bridge as the error of a two-point interpolation for Wiener process. The same fact is essentially true if we replace the interval $[0,1]$ with arbitrary interval $[t_0, t_1]$. Define interpolation error process by

$$\Delta(t) := W(t) - \left[\frac{t_1 - t}{t_1 - t_0}W(t_0) + \frac{t - t_0}{t_1 - t_0}W(t_1)\right], \qquad t_0 \le t \le t_1.$$

To see that Δ is a scaled Brownian bridge, we use that W has stationary increments and is $\frac{1}{2}$-self-similar. By combining two properties, we obtain that $\widetilde{W}(s) := \frac{W(t_0 + s(t_1 - t_0)) - W(t_0)}{\sqrt{t_1 - t_0}}$ is a Wiener process. It follows that

$$\Delta(t) = W(t) - W(t_0) - \frac{t - t_0}{t_1 - t_0}\left(W(t_1) - W(t_0)\right)$$

$$= \sqrt{t_1 - t_0}\left[\widetilde{W}\left(\frac{t - t_0}{t_1 - t_0}\right) - \frac{t - t_0}{t_1 - t_0}\widetilde{W}(1)\right]$$

$$= \sqrt{t_1 - t_0}\,\overset{\circ}{W}\left(\frac{t - t_0}{t_1 - t_0}\right),$$

where the Brownian bridge $\overset{\circ}{W}$ is related to the Wiener process $\widetilde{W}$ by (6.8). Using independence of increments of W, one can additionally show that if we take a set of disjoint intervals $\left([t_0^{(j)}, t_1^{(j)}]\right)_{1 \leq j \leq n}$ then the corresponding bridges $\Delta^{(j)}$ are independent.

We have shown how Brownian bridge emerges from a Wiener process. In opposite direction, Wiener process can be easily constructed on the base of Brownian bridge as follows. Let $\overset{\circ}{W}$ be a Brownian bridge and let X be an $\mathcal{N}(0,1)$-distributed random variable independent of $\overset{\circ}{W}$. Then the process

$$W(t) := \overset{\circ}{W}(t) + tX, \quad 0 \leq t \leq 1,$$

is a Wiener process restricted on $[0,1]$, since

$$\mathbb{E}\,W(s)W(t) = \mathbb{E}\left((\overset{\circ}{W}(s) + sX)(\overset{\circ}{W}(t) + tX)\right)$$

$$= \min\{s,t\} - st + st = \min\{s,t\} = K_W(s,t).$$

Finally, mention the Markov property of $\overset{\circ}{W}$. Let us take some instant $t_0 \geq 0$ as "the present time". Then we can represent "the future" of Brownian bridge $(\overset{\circ}{W}(t))_{t \geq t_0}$ as

$$\overset{\circ}{W}(t) = \frac{1 - t}{1 - t_0}\overset{\circ}{W}(t_0) + \left(\overset{\circ}{W}(t) - \frac{1 - t}{1 - t_0}\overset{\circ}{W}(t_0)\right),$$

where the fist term only depends on the present location of $\overset{\circ}{W}$, while the second term is independent from the past $(\overset{\circ}{W}(s))_{0 \leq s \leq t_0}$, because for all $s \leq t_0 \leq t$ we have

$$\mathbb{E}\,\overset{\circ}{W}(s)\left(\overset{\circ}{W}(t) - \frac{1 - t}{1 - t_0}\overset{\circ}{W}(t_0)\right) = s - st - \frac{1 - t}{1 - t_0}(s - st_0) = 0.$$

6.3 *Ornstein–Uhlenbeck process*

Ornstein–Uhlenbeck process is a Gaussian process $U(t), t \in \mathbb{R}$, satisfying assumptions

$$\mathbb{E}\, U(t) = 0, \qquad K_U(s,t) := \mathbb{E}\, U(s)U(t) = e^{-|s-t|/2}. \qquad (6.9)$$

This is an example of a Gaussian *stationary* process (see Section 5 for definition). In other words, for any $t_0 \in \mathbb{R}$ the process $X(t) := U(t_0 + t)$ is again Ornstein–Uhlenbeck process, since

$$\mathbb{E}\, X(s)X(t) = K_U(t_0 + s, t_0 + t) = e^{-|(t_0+s)-(t_0+t)|/2} = e^{-|s-t|/2}.$$

A typical trajectory of an Ornstein–Uhlenbeck process is represented on Figure 6.3. Its local irregular behavior is the same as that of Wiener process. On a bounded time interval, the only difference is the absence fixed zero points. On unbounded time intervals however the difference is substantial: the range of Wiener process increases as the square root of time while the range of Ornstein–Uhlenbeck process increases as the square root of logarithm of time.

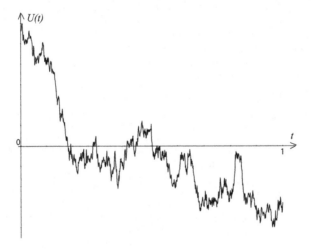

Fig. 6.3   Trajectory of an Ornstein–Uhlenbeck process.

There is a simple bilateral relation between Ornstein–Uhlenbeck process and Wiener process. If W is a Wiener process, then

$$U(t) := e^{-t/2}W(e^t) \qquad (6.10)$$

is an Ornstein–Uhlenbeck process, since

$$\mathbb{E}\, U(s)U(t) = e^{-(s+t)/2} K_W(e^s, e^t) = e^{-(s+t)/2} e^{\min\{s,t\}}$$
$$= e^{-\max\{s,t\}/2 + \min\{s,t\}/2} = e^{-|s-t|/2}.$$

Conversely, if U is an Ornstein–Uhlenbeck process, then

$$W(t) := \begin{cases} \sqrt{t}\, U(\ln t), & t > 0, \\ 0, & t = 0, \end{cases}$$

is a Wiener process.

Not surprisingly, Ornstein–Uhlenbeck process inherits the Markov property through this transform. Taking some instant t_0 as "the present time", we can represent "the future" of Ornstein–Uhlenbeck process $(U(t))_{t \geq t_0}$ as

$$U(t) = e^{-(t-t_0)/2} U(t_0) + \left(U(t) - e^{-(t-t_0)/2} U(t_0) \right), \tag{6.11}$$

where the fist term only depends on the present location of U, while the second term is independent from the past $(U(s))_{s \leq t_0}$, because for $s \leq t_0 \leq t$ we have

$$\mathbb{E}\, U(s) \left(U(t) - e^{-(t-t_0)/2} U(t_0) \right) = e^{-(t-s)/2} - e^{-(t-t_0)/2} e^{-(t_0-s)/2} = 0.$$

One can say that Ornstein–Uhlenbeck process has a very short memory, because the past-dependent part in (6.11) is decreasing exponentially fast as t goes to infinity.

6.4 *Fractional Brownian motion*

We introduce now a family of processes that include Wiener process and inherit most of its main properties except for independence of increments and Markov property. Let $H \in (0, 1]$ be a self-similarity parameter, or Hurst parameter. A *fractional Brownian motion* (fBm) with parameter H is a Gaussian process $B^H(t), t \in \mathbb{R}$, with zero mean and covariance function

$$K_H(s, t) := \frac{1}{2} \left(|s|^{2H} + |t|^{2H} - |s - t|^{2H} \right). \tag{6.12}$$

The word "fractional" in the process name hints at the relation between fBm and fractional integration, see [81].

Typical trajectories of fBm are represented on Figure 6.4. Notice that trajectories corresponding to a higher H are much more regular.

The origin of fBm is Kolmogorov's note [51], see further historical remarks in [72]. To a large extent, the importance of fBm became understood after Mandelbrot's works [103], especially [66].

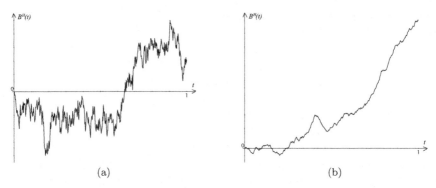

Fig. 6.4   Trajectories of fBm. (a) $H < 1/2$. (b) $H > 1/2$.

The most interesting cases are $H = 1/2$ and $H = 1$. For $H = 1/2$ we have

$$K_H(s,t) = \begin{cases} \min(|s|, |t|), & \text{if } st \geq 0, \\ 0, & \text{if } st \leq 0. \end{cases}$$

In other words, on the positive and negative half-lines we observe two independent Wiener processes.

For $H = 1$ we have $K_H(s,t) = st$, hence

$$\mathbb{V}ar(B^1(t) - tB^1(1)) = K_H(t,t) - 2tK_H(t,1) + t^2 K_H(1,1) = t^2 - 2t^2 + t^2 = 0.$$

In other words, $B^1(t) = tB^1(1)$ is a degenerated process with random linear sample paths.

Speaking about the family fBm at large, one may notice that the process increments are negatively dependent for $H < 1/2$, for $H = 1/2$ they are independent, and for $H > 1/2$ they are positively dependent. Moreover, the degree of this dependence is increasing with H and attains its maximum at $H = 1$, where the correlation coefficient for increments reaches 1.

Exercise 6.3. Let $0 \leq s_1 \leq t_1 \leq s_2 \leq t_2$. Prove that

$$\text{cov}\Big(B^H(t_1) - B^H(s_1), B^H(t_2) - B^H(s_2)\Big) \begin{cases} \geq 0, & \frac{1}{2} < H < 1, \\ \leq 0, & 0 < H < \frac{1}{2}. \end{cases}$$

Fractional Brownian motion plays an important role in limit theorems for random processes as well as the Wiener process, especially in the long range dependence case. In the subsequent studies of teletraffic systems we will only meet fBm with parameters $H \geq 1/2$, since the investigated processes by their nature have non-negatively dependent increments.

Let us discuss the basic properties of fBm as we did for Wiener process.

- *Alternative definition*: a Gaussian process $B^H(t), t \in \mathbb{R}$, is a fBm if and only if it satisfies

$$\mathbb{E} B^H(t) = 0, \qquad t \in \mathbb{R}; \qquad (6.13)$$

$$B^H(0) = 0; \qquad (6.14)$$

$$\mathbb{E} (B^H(t) - B^H(s))^2 = |s - t|^{2H}, \qquad s, t \in \mathbb{R}. \qquad (6.15)$$

Assume first that B^H is a fBm. Then (6.13) holds by definition. Since $\mathbb{E} B^H(0)^2 = K_H(0,0) = 0$, we obtain (6.14). Finally,

$$\mathbb{E} (B^H(s) - B^H(t))^2 = \mathbb{E} B^H(s)^2 + \mathbb{E} B^H(t)^2 - 2\mathbb{E} (B^H(s)B^H(t))$$
$$= |s|^{2H} + |t|^{2H} - \left(|s|^{2H} + |t|^{2H} - |s - t|^{2H} \right)$$
$$= |s - t|^{2H},$$

as required in (6.15). Conversely, let equalities (6.13)-(6.15) be true. Then by plugging $t = 0$ in (6.15) and using (6.14) we obtain $\mathbb{E} B^H(s)^2 = |s|^{2H}$. Similarly, $\mathbb{E} B^H(t)^2 = |t|^{2H}$. Finally,

$$\mathbb{E} (B^H(s)B^H(t)) = \frac{1}{2} \mathbb{E} \left(B^H(s)^2 + B^H(t)^2 - (B^H(s) - B^H(t))^2 \right)$$
$$= \frac{1}{2} \left(|s|^{2H} + |t|^{2H} - |s - t|^{2H} \right),$$

as required in the definition of fBm.

- *H-self-similarity*: for any $c > 0$ consider $Y(t) := B^H(ct)/c^H$. Then the process Y is also a fBm.

  Indeed, it is clear that Y inherits from B^H the property of being a Gaussian process. Moreover, we have $\mathbb{E} Y(t) = \frac{\mathbb{E} B^H(ct)}{c^H} = 0$ and

$$K_Y(s,t) := \mathbb{E} Y(s)Y(t) = \frac{\mathbb{E} (B^H(cs)B^H(ct))}{c^H c^H}$$
$$= \frac{1}{2c^{2H}} \left(|cs|^{2H} + |ct|^{2H} - |cs - ct|^{2H} \right)$$
$$= \frac{1}{2} \left(|s|^{2H} + |t|^{2H} - |s - t|^{2H} \right),$$

as required.

- *Stationary increments*: for any $t_0 \in \mathbb{R}$ the process

$$Y(t) := B^H(t_0 + t) - B^H(t_0)$$

is also a fBm; Again, it is clear that Y inherits from W the property of being a Gaussian process. Moreover, all features of alternative definition (6.13)-(6.15) are present. Indeed, we have

$$\mathbb{E}\left(Y(s) - Y(t)\right)^2 = \mathbb{E}\left(B^H(t_0 + s) - B^H(t_0 + t)\right)^2$$
$$= |(t_0 + s) - (t_0 + t)|^{2H} = |s - t|^{2H},$$

as required in (6.15).

Taken all together, the gaussianity, the stationarity of increments, and self-similarity uniquely define the class of fBM. This explains its importance for limit theorems in applied models. See Sections 13.1 and 13.3 for more explanations and examples.

We can also remark that the value of H is related not only to the self-similarity of fBm but also to the smoothness (Hölder property) of its sample paths.

For $H \in (\frac{1}{2}, 1]$ the increments of fBm have a long range dependence property. For general studies of this (not quite precisely defined) phenomenon, we refer to [29]; also see an intuitively clear example in Subsection 13.3.

It is just curious what happens with fBm when H goes to zero. Notice that

$$\lim_{H \to 0} K_H(s,t) = K_0(s,t) := \begin{cases} \frac{1}{2}, & s \neq t, \\ 1, & s = t. \end{cases}$$

One can interpret K_0 as a covariance of an extremely irregular process

$$W^0(t) := Y + Z(t), \qquad t \in \mathbb{R},$$

where Y and all $Z(t), t \in \mathbb{R}$, are independent $\mathcal{N}(0, \frac{1}{2})$-distributed random variables.

As in the case of Wiener process, a combination of wonderful properties of fractional Brownian motion is, essentially, unique, as the following proposition shows.

Proposition 6.4. *Let $X(t), t \geq 0$, be an H-self-similar Gaussian process with stationary increments. Then $H \leq 1$ and there exist $c \in \mathbb{R}$, $\sigma \geq 0$ and a a fractional Brownian motion B^H such that*

$$X(t) = \begin{cases} \sigma B^H(t), & t \geq 0, \qquad 0 < H < 1, \\ \sigma B^1(t) + ct, & t \geq 0, \quad H = 1. \end{cases} \tag{6.16}$$

Proof. We have already seen in (6.5) and (6.6) that self-similarity yields

$$\mathbb{E}\,X(k) = c\,k^H, \qquad k \geq 0,$$

and

$$\mathbb{V}arX(k) = \sigma^2 k^{2H}, \qquad k \geq 0,$$

with appropriate c and σ, as well as $X(0) = 0$. Then, stationarity of increments yields for any $0 \leq s \leq t$

$$c\,(t^H - s^H) = \mathbb{E}\,(X(t) - X(s)) = \mathbb{E}\,X(t - s) = c\,(t - s)^H.$$

If $H \neq 1$, this is only possible for $c = 0$.

Next, using again the stationarity of increments, we obtain

$$\mathbb{V}ar(X(t) - X(s)) = \mathbb{V}arX(t - s) = \sigma^2(t - s)^{2H}.$$

On the other hand,

$$\mathbb{V}ar(X(t) - X(s)) = \mathbb{V}arX(s) + \mathbb{V}arX(t) - 2\mathrm{cov}(X(s), X(t))$$
$$= \sigma^2 s^{2H} + \sigma^2 t^{2H} - 2\mathrm{cov}(X(s), X(t)).$$

By comparing the two expressions,

$$\mathrm{cov}(X(s), X(t)) = \frac{\sigma^2}{2}\left(s^{2H} + t^{2H} - (t - s)^{2H}\right),$$

and it becomes clear that (dropping out the trivial case $\sigma = 0$) the process

$$B^H(t) := \sigma^{-1}(X(t) - ct^H)$$

is a fractional Brownian motion that satisfies (6.16). $\qquad\square$

6.5 *Brownian sheet*

A Gaussian process $W^C(t), t \in \mathbb{R}_+^d$, is called *Brownian sheet* or *Wiener–Chentsov field*, if

$$\mathbb{E}\,W^C(t) = 0, \qquad \mathbb{E}\,W^C(s)W^C(t) = \prod_{l=1}^{d}\min\{s_l, t_l\}. \tag{6.17}$$

For $d = 1$ we obtain a classical Wiener process.

Covariance of Brownian sheet has a transparent geometric meaning. Let us relate to any point $t \in \mathbb{R}_+^d$ a parallelepiped

$$[0, t] := \{u \in \mathbb{R}^d : 0 \leq u_l \leq t_l, 1 \leq l \leq d\}.$$

Then we have

$$\prod_{l=1}^{d} \min\{s_l, t_l\} = \lambda^d \left([0,s] \cap [0,t]\right),$$

where λ^d is the Lebesgue measure in $\mathbb{R}_+^d$.

It follows from (6.17) that $W^C(t)$ is H-self-similar for $H = \frac{d}{2}$. Namely, for any $c > 0$ consider $Y(t) := \frac{W^C(ct)}{c^H}$. Then Y is also a Brownian sheet.

Brownian sheet has a certain property of "independent increments" extending that of Wiener process. Let us start with the simpler case $d = 2$. Take $s = (s_1, s_2)$, $t = (t_1, t_2)$ in $\mathbb{R}_+^2$ and assume that $s_1 \leq t_1$, $s_2 \leq t_2$. The *increment* associated to the rectangle

$$[s,t] := \{u \in \mathbb{R}^2 : s_1 \leq u_1 \leq t_1, \, s_2 \leq u_2 \leq t_2\},$$

is defined by the formula

$$\Delta_{s,t}(W^C) := W^C(t_1, t_2) - W^C(s_1, t_2) - W^C(t_1, s_2) + W^C(s_1, s_2).$$

For larger dimensions d the construction goes in the same way. Take $s, t \in \mathbb{R}_+^d$ and assume that $s_l \leq t_l$, $1 \leq l \leq d$. The *increment* associated to the rectangle

$$[s,t] := \{u \in \mathbb{R}^d : s_l \leq u_l \leq t_l, \, 1 \leq l \leq d\},$$

is defined by

$$\Delta_{s,t}(W^C) := \sum_{Q} (-1)^{\#(Q)-n} W^C(v^Q).$$

Here the sum is taken over all 2^d subsets Q of $[1..d]$, $\#(Q)$ denotes the number of elements in Q and v^Q is a vertex of rectangle $[s,t]$ defined by

$$v_i^Q := \begin{cases} t_i, & i \in Q, \\ s_i, & i \notin Q. \end{cases}$$

The independence of increments can be now stated as follows.

Proposition 6.5. *Let $s^{(j)}, t^{(j)} \in \mathbb{R}_+^d$, $1 \leq j \leq n$, be as above. Assume that the rectangles $[s^{(j)}, t^{(j)}]$ have no common interior points. Then the increments $\Delta_{s^{(j)}, t^{(j)}}(W^C)$ are independent.*

Brownian sheet is a special case of *tensor product* of random processes, which is a random field with covariance

$$K(s,t) = \prod_{l=1}^{d} K_l(s_l, t_l),$$

where $K_l(\cdot, \cdot)$ are covariance functions of one-parametric processes that do not necessarily coincide.

6.6 Lévy's Brownian function

A Gaussian process $W^L(t), t \in \mathbb{R}^d$, is called *Lévy's Brownian function* or *Lévy field* if

$$\mathbb{E}\, W^L(t) = 0, \qquad \mathbb{E}\, W^L(s)W^L(t) = \frac{1}{2}\left(||s|| + ||t|| - ||s - t||\right). \qquad (6.18)$$

Here $|| \cdot ||$ denotes Euclidean norm in $\mathbb{R}^d$. This process was introduced and explored by P.Lévy in [58].

For $d = 1$ Lévy's Brownian function reduces to a couple of independent Wiener processes (for $t \geq 0$ and for $t \leq 0$).

Exactly as in the one-parametric case, there exists an equivalent alternative definition: a Gaussian process $W^L(t), t \in \mathbb{R}^d$, is a Lévy's Brownian function if and only if it satisfies

$$\mathbb{E}\, W^L(t) = 0, \qquad t \in \mathbb{R}^d;$$
$$W^L(0) = 0;$$
$$\mathbb{E}\, (W^L(s) - W^L(t))^2 = ||s - t||, \qquad s, t \in \mathbb{R}^d. \qquad (6.19)$$

It follows from (6.18) that $W^L(t)$ is $\frac{1}{2}$-self-similar. For any $c > 0$ let $Y(t) := \frac{W^L(ct)}{c^{1/2}}$. Then the process Y is also a Lévy's Brownian function.

The process W^L is also rotation invariant. If $U : \mathbb{R}^d \mapsto \mathbb{R}^d$ is a linear isometry, then the process $Y(t) := W^L(Ut)$ is also a Lévy's Brownian function, because

$$K_Y(s,t) = \frac{1}{2}\left(||Us|| + ||Ut|| - ||U(s - t)||\right) = \frac{1}{2}\left(||s|| + ||t|| - ||s - t||\right).$$

6.7 Further extensions

Brownian sheet and Lévy's Brownian function extend the notion of Wiener process to the case of d-parametric random fields, each in its own fashion. Similar extensions are equally possible for H-fractional Brownian motion with arbitrary $H \in (0, 1]$.

Fractional Brownian sheet $W^H(t), t \in \mathbb{R}^d$, is defined as a Gaussian process with zero mean and tensor product covariance

$$\mathbb{E}\, W^H(s)W^H(t) = 2^{-d}\prod_{l=1}^{d}\left(|s_l|^{2H} + |t_l|^{2H} - |s_l - t_l|^{2H}\right).$$

One can go even further by letting Hurst parameter H depend on the coordinate.

Fractional Lévy's Brownian function $W^{L,H}(t), t \in \mathbb{R}^d$, is defined as a Gaussian process with zero mean and covariance

$$\mathbb{E}\, W^{L,H}(s)W^{L,H}(t) = \frac{1}{2}\left(||s||^{2H} + ||t||^{2H} - ||s-t||^{2H}\right).$$

The reader is invited to explore the self-similarity properties of these processes as an exercise.

Inspired by (6.18), one can imagine the following "abstract" extension of Lévy's Brownian function. Let (T, ρ) be a metric space with a marked point t_0. We call *Lévy's Brownian function on T* a Gaussian process $W^L(t), t \in T$, with zero mean and covariance

$$K(s,t) := \mathbb{E}\, W^L(s)W^L(t) = \frac{1}{2}\left(\rho(s,t_0) + \rho(t,t_0) - \rho(s,t)\right).$$

It is a hard problem to determine for which metric spaces (T, ρ) such a process exists, or, in other words, when the function $K(\cdot, \cdot)$ is non-negative definite. The answer is positive for Euclidean spaces, for spheres equipped with the geodesic distance, for hyperbolic spaces and in few other cases. In $\mathbb{R}^2$, *any* norm generates a distance such that Lévy's Brownian function exist. However, already in $\mathbb{R}^3$ one can construct a norm generating a distance such that the corresponding Lévy's Brownian function does not exist. See further examples of Lévy's Brownian functions and their fractional generalizations on spheres, hyperbolic spaces, etc. in [17, 20, 21, 33, 37, 42, 43, 97, 98].

Our next object is a fractional extension of Ornstein–Uhlenbeck process. Actually, there are several ways for such extension leading to different processes. Looking at the covariance formula (6.9), we may define the *fractional Ornstein–Uhlenbeck process* U^H with $H \in (0,1)$ as a Gaussian process on the real line satisfying

$$\mathbb{E}\, U^H(t) = 0, \qquad K_{U^H}(s,t) := \mathbb{E}\, U^H(s)U^H(t) = e^{-|s-t|^{2H}}.$$

On the other hand, looking at the representation via Wiener process (6.10), we should define the same object by

$$\widetilde{U}^H := e^{-tH}B^H(e^t).$$

This leads to covariance

$$\mathbb{E}\, \widetilde{U}^H(s)\widetilde{U}^H(t) = e^{-(t+s)H}\left(e^{2sH} + e^{2tH} - \left|e^s - e^t\right|^{2H}\right)/2$$

$$= \left(e^{(s-t)H} + e^{(t-s)H} - \left|e^{(s-t)/2} - e^{(t-s)/2}\right|^{2H}\right)/2$$

$$= \left(e^{|s-t|H} + e^{-|s-t|H} - \left(e^{|s-t|/2} - e^{-|s-t|/2}\right)^{2H}\right)/2.$$

Both U^H and $\widetilde{U}^H$ are centered Gaussian stationary processes but they are clearly different. For further extensions and applications in telecommunication models, we refer to [112].

Houdré and Villa introduced in [40] another interesting extension of fBm. A Gaussian process $W^{H,K}(t), t \in \mathbb{R}_+$, is called (H, K)- *bifractional Brownian motion* if

$$\mathbb{E}\, W^{H,K}(t) = 0, \quad \mathbb{E}\, W^{H,K}(s) W^{H,K}(t) = \frac{1}{2^K} \left((t^{2H} + s^{2H})^K - |t - s|^{2HK} \right).$$

Note that letting $K = 1$ yields a usual fractional Brownian motion W^H. The process $W^{H,K}$ exists provided that $0 < H \leq 1$, $0 < K \leq 2$, and $HK \leq 1$.

For further information about this process we refer to [6, 56, 104]. As for applications, Marouby [69] showed that $W^{H,K}$ with $H = \frac{1}{2}$ appears as a limit in a version of Mandelbrot's "micropulse" model. We discuss it later in Section 14.

We refer to [1, 12, 60, 61] for further information about Gaussian processes and distributions.

7   Random Measures and Stochastic Integrals

7.1   *Random measures with uncorrelated values*

Let $(\mathcal{R}, \mu)$ be a measure space and $\mathbf{A} = \{A \subset \mathcal{R} : \mu(A) < \infty\}$. A family of random variables $\{X(A), A \in \mathbf{A}\}$ is called a centered *random measure with uncorrelated values*, if

$$\mathbb{E}\, X(A) = 0, \qquad A \in \mathbf{A},$$

and

$$\mathrm{cov}(X(A_1), X(A_2)) = \mu(A_1 \cap A_2), \qquad A_1, A_2 \in \mathbf{A}. \tag{7.1}$$

For disjoint A_1, A_2 we have $\mathrm{cov}(X(A_1), X(A_2)) = 0$, i.e. $X(A_1)$ and $X(A_2)$ are uncorrelated, which explains the name for X.

The measure μ is called *intensity measure for X*.

Notice that if the sets $A_1, A_2, \ldots$ are disjoint and $A := \cup_j A_j$ then by (7.1)

$$\mathbb{E}\left[X(A) - \sum_j X(A_j)\right]^2$$

$$= \mathbb{E}\,X(A)^2 + \sum_j \mathbb{E}\,X(A_j)^2 - 2\sum_j \mathbb{E}\,X(A)X(A_j)$$

$$-2\sum_{1\leq i<j} \mathbb{E}\,X(A_i)X(A_j)$$

$$= \text{cov}(X(A),X(A)) + \sum_j \text{cov}(X(A_j),X(A_j)) - 2\sum_j \text{cov}(X(A),X(A_j))$$

$$-2\sum_{i<j}^m \text{cov}(X(A_i),X(A_j))$$

$$= \mu(A) + \sum_j \mu(A_j) - 2\sum_j \mu(A_j)$$

$$= \mu(A) + \mu(A) - 2\mu(A) = 0.$$

Therefore, we have an additivity property:

$$X(A) = \sum_j X(A_j) \qquad \text{almost surely.}$$

This does not mean, however, that $X(\cdot)$ is a random *measure*. Notice that a set of probability zero where additivity may fail depends on the sets $A_1, A_2, \ldots$. Therefore, we are not allowed to claim that additivity holds with probability one for all ensembles $\{A_1, A_2, \ldots\}$ simultaneously, as required in the definition of a measure.

We will define now the *integrals* of the functions $f : \mathcal{R} \mapsto \mathbb{R}$ with respect to a random measure X. For a step function $f := \sum_{j=1}^m c_j \mathbf{1}_{\{A_j\}}$ with $A_j \in \mathbf{A}$, the integral is naturally defined as

$$I_f := \int_{\mathcal{R}} f\,dX := \sum_{j=1}^m c_j X(A_j).$$

Notice that for any two step functions, f as above and $g := \sum_{i=1}^n d_i \mathbf{1}_{\{B_i\}}$

we have

$$
\begin{aligned}
\mathbb{E}\left(I_f I_g\right) &= \sum_{j=1}^{m} \sum_{i=1}^{n} c_j d_i \,\mathbb{E}\left(X(A_j)X(B_i)\right) \\
&= \sum_{j=1}^{m} \sum_{i=1}^{n} c_j d_i \,\mu(A_j \cap B_i) \\
&= \sum_{j=1}^{m} \sum_{i=1}^{n} c_j d_i \int_{\mathcal{R}} \mathbf{1}_{\{A_j\}} \mathbf{1}_{\{B_i\}} d\mu \\
&= \int_{\mathcal{R}} \left(\sum_{j=1}^{m} c_j \mathbf{1}_{\{A_j\}}\right) \left(\sum_{i=1}^{n} d_i \mathbf{1}_{\{B_i\}}\right) d\mu \\
&= \int_{\mathcal{R}} f g \, d\mu.
\end{aligned}
\tag{7.2}
$$

It follows that

$$
\begin{aligned}
\mathbb{E}\left(I_f - I_g\right)^2 &= \mathbb{E}\,I_f^2 + \mathbb{E}\,I_g^2 - 2\mathbb{E}\left(I_f I_g\right) \\
&= \int_{\mathcal{R}} f^2 d\mu + \int_{\mathcal{R}} g^2 d\mu - 2 \int_{\mathcal{R}} f g \, d\mu \\
&= \int_{\mathcal{R}} (f - g)^2 d\mu.
\end{aligned}
\tag{7.3}
$$

In particular, if $f = g$ are two different representations of the same step function, we have $\mathbb{E}\left(I_f - I_g\right)^2 = 0$, i.e. $I_f = I_g$ almost surely. This means that our definition of the integral $\int_{\mathcal{R}} f dX$ is correct: it does not depend of a specific representation of f as a sum.

The moment properties of integrals are as follows. Clearly,

$$
\mathbb{E} \int_{\mathcal{R}} f dX = \sum_{j=1}^{m} c_j \mathbb{E}\,X(A_j) = 0
\tag{7.4}
$$

and by (7.2)

$$
\mathbb{V}ar \int_{\mathcal{R}} f dX = \mathbb{E}\,I_f^2 = \int_{\mathcal{R}} f^2 d\mu.
\tag{7.5}
$$

Notice that the integral is a linear functional of the integrand, i.e.

$$
\int_{\mathcal{R}} (cf) dX = c \int_{\mathcal{R}} f dX,
\tag{7.6}
$$

$$
\int_{\mathcal{R}} (f + g) dX = \int_{\mathcal{R}} f dX + \int_{\mathcal{R}} g dX.
\tag{7.7}
$$

Here the first property is trivial. The second one also becomes obvious, when we represent f and g by means of the same collection of step sets A_j (which is always possible).

Now we extend the notion of integral to all functions

$$f \in \mathbb{L}_2(\mathcal{R}, \mu) := \left\{ f : \mathcal{R} \mapsto \mathbb{R} : \int_\mathcal{R} |f|^2 d\mu < \infty \right\}.$$

Recall that any function of this type admits an approximation in $\mathbb{L}_2(\mathcal{R}, \mu)$ by step functions, e.g. for

$$f_n(u) = \sum_{j=-n^2}^{n^2} \frac{j}{n} \mathbf{1}_{\{\frac{j}{n} \leq f(u) < \frac{j+1}{n}\}}(u)$$

we have

$$\int_\mathcal{R} |f_n - f|^2 d\mu \to 0,$$

which is also denoted by $f_n \overset{\mathbb{L}_2}{\to} f$. So take _arbitrary_ sequence of step functions f_n such that $f_n \overset{\mathbb{L}_2}{\to} f$. Notice that

$$\mathbb{E}\left(I_{f_n} - I_{f_m}\right)^2 = \int_\mathcal{R} (f_n - f_m)^2 d\mu \to 0, \qquad \text{as } n, m \to \infty.$$

Therefore, the sequence of random variables (I_{f_n}) is a Cauchy sequence in $\mathbb{L}_2$ on the corresponding probability space. Any $\mathbb{L}_2$-space is complete, hence any Cauchy sequence is convergent. We conclude that there exist a limit I_f such that

$$\|I_{f_n} - I_f\|_{\mathbb{L}_2} = \mathbb{E}\left(I_{f_n} - I_f\right)^2 \to 0, \qquad \text{as } n \to \infty.$$

Notice that the limit does not depend on the choice of converging sequence (f_n). Indeed, if f_n and g_n are two sequences converging to the same limit f, we have

$$\|I_{f_n} - I_{g_n}\|_{\mathbb{L}_2} = \|I_{f_n - g_n}\|_{\mathbb{L}_2} = \|f_n - g_n\|_{\mathbb{L}_2} \to 0, \qquad \text{as } n \to \infty.$$

Hence the limit of I_{f_n} and the limit of I_{g_n} coincide.

Therefore, we may define

$$\int_\mathcal{R} f dX := I_f,$$

without taking care about particular choice of an approximating sequence. All properties and formulas mentioned above for the integrals of step functions will be still verified for this extended version. We leave the easy proof of this fact to the reader.

In conclusion, notice that equalities (7.2), (7.3)

$$\mathbb{E}\left(I_f I_g\right) = \int_{\mathcal{R}} f g \, d\mu \qquad (7.8)$$

and

$$\mathbb{E}\left(I_f - I_g\right)^2 = \int_{\mathcal{R}} (f - g)^2 d\mu \qquad (7.9)$$

have a clear geometric meaning: integration preserves the scalar products of integrands and the square mean distances between them. Therefore, (7.8) and (7.9) are referred to as versions of the *isometric property* of the integral.

Complex measures with uncorrelated values

Complex-valued measures with uncorrelated values and the respective integrals are defined along the same lines as their real analogues. Instead of (7.1) the covariance of complex measure is defined by

$$\operatorname{cov} X(A)\overline{X(B)} = \mu(A \cap B).$$

Here the variables $X(\cdot) \in \mathbb{C}$ and the integrands f in $\int_{\mathcal{R}} f dX$ are complex-valued functions $f \in \mathbb{L}_{2,\mathbb{C}}(\mathcal{R}, \mathcal{A}, \mu)$. Isometric property (7.8) reads now as

$$\mathbb{E}\left(\int_{\mathcal{R}} f dX \cdot \overline{\int_{\mathcal{R}} g dX}\right) = \int_{\mathcal{R}} f \, \overline{g} \, d\mu. \qquad (7.10)$$

Complex integration is necessary for spectral representation of a stationary process

$$Y(t) = \int_{-\infty}^{\infty} e^{itu} dX(u),$$

where the corresponding intensity measure μ is the spectral measure of process Y, see details in Section 10. Even if Y is a real-valued process, the corresponding random measure X typically is complex-valued.

7.2 Gaussian white noise

Let $(\mathcal{R}, \mathcal{A}, \mu)$ be a measure space. Let as before $\mathbf{A} = \{A \in \mathcal{A} : \mu(A) < \infty\}$. A random process $\{\mathcal{W}(A), A \in \mathbf{A}\}$ is called a *Gaussian white noise* with intensity measure μ, if $\mathcal{W}$ is a centered random measure with uncorrelated values and a Gaussian process, i.e. all finite-dimensional distributions of

$\mathcal{W}$ are Gaussian. By the definition, the expectation and covariance for the Gaussian white noise are

$$\mathbb{E}\,\mathcal{W}(A) = 0,$$

$$\text{cov}(\mathcal{W}(A), \mathcal{W}(B)) = \mathbb{E}\,\mathcal{W}(A)\mathcal{W}(B) = \mu(A \cap B).$$

As we know, these two formulae determine all finite-dimensional distributions of $\mathcal{W}$.

According to the previous general setting, the integral over white noise

$$\int_{\mathcal{R}} f d\mathcal{W} \tag{7.11}$$

is well defined for any $f \in \mathbb{L}_2(\mathcal{R}, \mu)$. It is called Gaussian *white noise integral*.

As shown below, many Gaussian processes admit a convenient definition or representation by means of Gaussian white noise integral.

Notice that any integral (7.11) is a normal random variable. Indeed, if f is a step function, then by definition the integral is a linear combination of components of a Gaussian random vector. Since arbitrary function $f \in \mathbb{L}_2$ admits an approximation in $\mathbb{L}_2$ by step functions, we conclude by isometric property that the integral (7.11) is an $\mathbb{L}_2$-limit of normal random variables. Furthermore, since $\mathbb{L}_2$-convergence implies convergence in distribution, and the set of normal distributions is closed w.r.t. to convergence in distribution, we conclude that the integral is normally distributed in the general case, too. Recall that we know from (7.4) and (7.5) the expectation and the variance of this integral. Therefore, the distribution of (7.11) is $\mathcal{N}\left(0, \int_{\mathcal{R}} f^2 d\mu\right)$.

In addition to general properties of centered measures with uncorrelated values described in the previous subsection, a Gaussian white noise has a supplementary feature: if the sets $A_1, \ldots, A_n$ are disjoint, then the variables $\mathcal{W}(A_1), \ldots, \mathcal{W}(A_n)$ are *independent*. Indeed, these variables form a Gaussian vector and they are uncorrelated. Hence, see Corollary 4.11, they are independent.

Thus we are led to the following definition.

Let $\mathcal{R}$ be a set equipped with an algebra of its subsets $\mathbf{A}$. A random process $\{X(A), A \in \mathbf{A}\}$ is called an *independently scattered measure* or a *random measure with independent values* if for any disjoint sets $A_1, \ldots, A_n$ in $\mathbf{A}$ the variables $X(A_1), \ldots, X(A_n)$ are independent and

$$X\left(\bigcup_{j=1}^{n} A_j\right) = \sum_{j=1}^{n} X(A_j) \qquad \text{almost surely.}$$

We see that a Gaussian white noise is a first example of independently scattered measure. Other examples will follow, such as a centered Poisson measure and stable measures. Notice that, unlike for a measure with uncorrelated values, the values of independently scattered measure need not have finite variance.

7.3 *Integral representations*

The properties of any centered Gaussian process $X(t), t \in T$, are entirely determined by its covariance function $K_X(s,t) = \mathbb{E}\, X(s)X(t)$. To construct such a process, it is enough to have a Gaussian white noise $\mathcal{W}$ on a measure space $(\mathcal{R}, \mathcal{A}, \mu)$ and a system of functions $\{m_t, t \in T\} \subset \mathbb{L}_2(\mathcal{R}, \mathcal{A}, \mu)$ such that

$$(m_s, m_t) := \int_{\mathcal{R}} m_s(u)m_t(u)\, \mu(du) = K_X(s,t), \qquad s, t \in T.$$

In this case the process

$$\widetilde{X}(t) := \int_{\mathcal{R}} m_t\, d\mathcal{W}, \qquad t \in T, \tag{7.12}$$

has a required covariance $K_X(s,t)$. We call expression (7.12) an *integral representation* of X.

7.3.1 *Wiener process*

We set $(\mathcal{R}, \mathcal{A}, \mu) := (\mathbb{R}_+, \mathcal{B}, \lambda)$, where $\mathcal{B}$ is Borel σ-field, λ is Lebesgue measure and

$$m_t(u) := \mathbf{1}_{[0,t]}(u).$$

Obviously,

$$\widetilde{X}(t) = \int \mathbf{1}_{[0,t]}(u)d\mathcal{W}(u) = \mathcal{W}([0,t])$$

is a Wiener process, since

$$K_{\widetilde{X}}(s,t) = \mathbb{E}\, \mathcal{W}([0,s])\mathcal{W}([0,t]) = \lambda([0,s] \cap [0,t])$$
$$= \lambda([0, \min\{s,t\}]) = \min\{s,t\}.$$

Conversely, if a Wiener process W is given, we may obtain the related white noise $\mathcal{W}$ by defining it first on the unions of disjoint intervals as

$$\mathcal{W}\left(\bigcup_{j=1}^{m}[s_j, t_j]\right) := \sum_{j=1}^{m}(W(t_j) - W(s_j)),$$

and then extending the definition to arbitrary Borel sets by density arguments. The integral over this white noise is then often called stochastic integral with respect to the Wiener process W. Thus, for any $f \in \mathbb{L}_2(\mathbb{R}, \lambda)$ one writes

$$\int_{\mathbb{R}} f dW := \int_{\mathbb{R}} f d\mathcal{W}.$$

7.3.2   *Brownian bridge*

We give two representations for Brownian bridge $\overset{\circ}{W}$. The first one emerges from the representation for Wiener process through linear relation

$$\overset{\circ}{W}(t) = W(t) - tW(1).$$

Clearly, the functions

$$m_t^0(u) := (m_t - t m_1)(u) = (1 - t)\mathbf{1}_{[0,t]}(u) - t\,\mathbf{1}_{(t,1]}(u)$$

provide an integral representation for $\overset{\circ}{W}$.

An alternative representation is built on a square $[0,1]^2$ equipped with 2-dimensional Lebesgue measure. We let

$$\widetilde{m}_t^0(u) := \mathbf{1}_{[0,t]\times[0,1-t]}(u).$$

Then (draw a picture for better understanding!)

$$\begin{aligned}(\widetilde{m}_s^0, \widetilde{m}_t^0) &= \lambda^2\left([0,s]\times[0,1-s]\bigcap[0,t]\times[0,1-t]\right)\\ &= \min(s,t)\cdot\min(1-s,1-t) = \min(s,t)\cdot(1-\max(s,t))\\ &= \min(s,t) - st,\end{aligned}$$

as required for a representation of $\overset{\circ}{W}$.

7.3.3   *Fractional Brownian motion*

One can construct an integral representation for H-fractional Brownian motion $B^H(t), t \in \mathbb{R}$, as follows [66]. Let $\mathcal{R} := \mathbb{R}$, $\mu := \lambda$ Lebesgue measure, and let $\mathcal{W}$ be a corresponding Gaussian white noise. Consider a process

$$\begin{aligned}B^H(t)\\ := \int_{-\infty}^{\infty} c_H \left((t-u)^{H-1/2}\mathbf{1}_{\{u\le t\}} - (-u)^{H-1/2}\mathbf{1}_{\{u\le 0\}}\right) d\mathcal{W}(u).\end{aligned} \quad (7.13)$$

Note that the integral is correctly defined exactly for $0 < H < 1$. If the normalizing factor c_H is chosen appropriately, we obtain an H-fractional Brownian motion since for all $t \geq s$

$$\mathbb{E}\left(B^H(t) - B^H(s)\right)^2$$

$$= c_H^2 \int_{-\infty}^{\infty} \left((t-u)^{H-1/2}\mathbf{1}_{\{u \leq t\}} - (s-u)^{H-1/2}\mathbf{1}_{\{u \leq s\}}\right)^2 du$$

$$\stackrel{u=s+v}{=} c_H^2 \int_{-\infty}^{\infty} \left((t-s-v)^{H-1/2}\mathbf{1}_{\{v \leq t-s\}} - (-v)^{H-1/2}\mathbf{1}_{\{v \leq 0\}}\right)^2 dv$$

$$\stackrel{v=(t-s)w}{=} c_H^2 (t-s)^{2H} \int_{-\infty}^{\infty} \left(1-w)^{H-1/2}\mathbf{1}_{\{w \leq 1\}} - (-w)^{H-1/2}\mathbf{1}_{\{w \leq 0\}}\right)^2 dw$$

$$= const \cdot (t-s)^{2H}.$$

By computing the integral, one can show that the required relation

$$\mathbb{E}\left(B^H(t) - B^H(s)\right)^2 = (t-s)^{2H}$$

is attained for

$$c_H = \frac{[\sin(\pi H)\Gamma(2H+1)]^{1/2}}{\Gamma(H+1/2)}.$$

7.3.4 *Riemann–Liouville processes and operators*

Opposite to previous and to subsequent examples, Riemann–Liouville processes are defined via their integral representation and not via covariance.

Recall that Riemann–Liouville fractional integration operator is defined by

$$R_\alpha f(t) := \frac{1}{\Gamma(\alpha)} \int_0^t (t-u)^{\alpha-1} f(u)du, \qquad \alpha > 0. \qquad (7.14)$$

Here

$$R_\alpha : \mathbb{L}_2[0,1] \mapsto \begin{cases} \mathbb{L}_p[0,1], & \alpha > \frac{1}{2} - \frac{1}{p}, \\ \mathbb{C}[0,1], & \alpha > \frac{1}{2}. \end{cases}$$

If $\alpha = 1$, we obtain the conventional integration operator. Riemann–Liouville operators have a remarkable semi-group property $R_\alpha R_\beta = R_{\alpha+\beta}$.

Similarly to (7.14), α-*Riemann–Liouville process* with $\alpha > 1/2$ is defined as a white noise integral on the real line (with Lebesgue measure as intensity measure)

$$R^\alpha(t) := \frac{1}{\Gamma(\alpha)} \int_0^t (t-u)^{\alpha-1} d\mathcal{W}(u), \qquad \alpha > 1/2.$$

When $\alpha = 1$, it coincides with a Wiener process. The restriction $\alpha > 1/2$ is necessary for correctness of the integral definition. In other words, the integration kernel $(t - \cdot)^{\alpha-1}$ must belong to $\mathbb{L}_2[0, t]$.

The process R^α is H-self-similar with index $H = 2\alpha - 1$.

The semi-group property yields $R_\alpha R^\beta = R^{\alpha+\beta}$.

As one can observe from (7.13), for $\alpha \in (1/2, 3/2)$ the local properties of α-Riemann–Liouville process are close to those of H-fractional Brownian motion with $H = \alpha - 1/2$.

However, opposite to the case of fractional Brownian motions, the family of Riemann–Liouville processes has no limitations in sample path smoothness, because there is no upper bound for parameter α.

Unlike fBm, the process R^α is not a process with stationary increments. As a compensation, it has another property called extrapolation homogeneity [62]: for any fixed $t_0 \geq 0$ the process

$$R^\alpha(t_0 + \cdot) - \mathbb{E}\left(R^\alpha(t_0 + \cdot)\big|\mathcal{F}_{t_0}\right)$$

has the same finite dimensional distributions as $R^\alpha(\cdot)$. Here $\mathcal{F}_{t_0}$ stands for the σ-field of the past for initial white noise $\mathcal{W}$ prior to time t_0, i.e.

$$\mathcal{F}_{t_0} = \sigma\{\mathcal{W}(A), A \subset [0, t_0]\}.$$

In econometric literature, where Riemann–Liouville process appears as a limit of discrete schemes, it is also often called "fractional Brownian motion". Therefore, one should thoroughly avoid a confusion with the "true" fractional Brownian motion defined by covariance (6.12). See [68] for further comparison of two processes.

7.3.5 *Brownian sheet*

This is a multi-parametric extension of the integral representation for Wiener process. We let here $(\mathcal{R}, \mathcal{A}, \mu) := (\mathbb{R}^d_+, \mathcal{B}^d, \lambda^d)$, where $\mathcal{B}^d$ is Borel σ-field on $\mathbb{R}^d$, λ^d is d-dimensional Lebesgue measure, and define the "rectangles"

$$[0, t] := \{u \in \mathcal{R} : 0 \leq u_j \leq t_j, \ 1 \leq j \leq d\}.$$

Then the intersection of rectangles is again a rectangle (draw a picture):

$$[0, s] \cap [0, t] = \{u \in \mathcal{R} : 0 \leq u_j \leq \min(s_j, t_j), \ 1 \leq j \leq d\}.$$

Therefore, the functions

$$m_t(u) := \mathbf{1}_{[0,t]}(u)$$

have a property

$$(m_s, m_t) = \lambda^d([0, s] \cap [0, t]) = \prod_{j=1}^{d} \min(s_j, t_j),$$

thus

$$W^C(t) := \int \mathbf{1}_{[0,t]}(u) d\mathcal{W}(u) = \mathcal{W}([0, t])$$

is a Brownian sheet.

7.3.6 *Lévy's Brownian function*

Recall that Lévy's Brownian function on $\mathbb{R}^d$ is defined by (6.18). We build now its white noise representation called *Chentsov integral-geometric construction* [16]. Let $\mathcal{R}$ be the space of all hyperplanes in $\mathbb{R}^d$. There exists a unique (up to a constant factor) measure μ on $\mathcal{R}$ that is invariant with respect to all unitary transformations of $\mathbb{R}^d$. For $t \in \mathbb{R}^d$ let A_t denote the set of all hyperplanes crossing the segment $\overline{0, t} := \{rt, 0 \leq r \leq 1\}$. It is easy to observe that $\mu(A_t)$ is proportional to the length of $\overline{0, t}$. Let the measure μ be normalized so that $\mu(A_t) = ||t||$. Let now $\mathcal{W}$ be a Gaussian white noise on $\mathcal{R}$ with intensity measure μ. Then

$$W^L(t) := \mathcal{W}(A_t) = \int_{\mathcal{R}} \mathbf{1}_{A_t} d\mathcal{W}, \qquad t \in \mathbb{R}^d,$$

is a Lévy's Brownian function on $\mathbb{R}^d$. Indeed, the relations $\mathbb{E}\, W^L(t) = 0$ and $W^L(0) = 0$ are obvious. Moreover, for any $s, t \in \mathbb{R}^d$ the symmetric difference

$$A_s \,\Delta\, A_t := (A_s \backslash A_t) \cup (A_t \backslash A_s)$$

consists of the hyperplanes crossing the segment

$$\overline{s, t} := \{s + r(t - s), 0 \leq r \leq 1\}$$

(we ignore the μ-null set of hyperplanes that contain one of the points $0, s, t$). Therefore,

$$\mathbb{E}\,(W^L(s) - W^L(t))^2 = \mathbb{E}\,(\mathcal{W}(A_s) - \mathcal{W}(A_t))^2$$
$$= \int_{\mathcal{R}} (\mathbf{1}_{\{A_s\}} - \mathbf{1}_{\{A_t\}})^2 d\mu = \mu(A_s \Delta A_t) = ||s - t||,$$

as required in (6.19).

One may "pack" this integral-geometric construction into $\mathbb{R}^d$, rendering it more elementary although less transparent. Indeed, let $\mathcal{R}_0$ be the set of all hyperplanes containing the origin. Then there exists a natural bijection between the sets $\mathcal{R} \backslash \mathcal{R}_0$ and $\mathbb{R}^d \backslash \{0\}$, i.e. to each hyperplane corresponds its point having the minimal distance to the origin. This bijection transforms the set A_t into the ball $\widetilde{A}_t$ of radius $||t||/2$ centered at $\frac{t}{2}$ (the segment $\overline{0,t}$ is a diameter for this ball; the reader is invited to check the correspondence between A_t and $\widetilde{A}_t$). The image of μ is a spherically symmetric measure $\widetilde{\mu} = dr\nu(d\theta)$, where $\nu(d\theta)$ is appropriately normalized uniform measure on the unit sphere (prove it!). It is clear that the measure $\widetilde{\mu}$ is different from Lebesgue measure on $\mathbb{R}^d$. If $\widetilde{\mathcal{W}}$ is a white noise on $\mathbb{R}^d$ with intensity measure $\widetilde{\mu}$, then $\widetilde{\mathcal{W}}(\widetilde{A}_t)$ is also a Lévy's Brownian function.

One can implement an integral-geometric construction, similar to Chentsov construction, for Lévy's Brownian function on a sphere (then, instead of hyperplanes, one should use the circles of maximal radius), on a hyperbolic space, and in some other cases.

7.4   *Poisson random measures and integrals*

Let $(\mathcal{R}, \mu)$ be a measure space and $\mathbf{A} = \{A \subset \mathcal{R}, \mu(A) < \infty\}$. A family of random variables $\{N(A), A \in \mathbf{A}\}$ is called *Poisson random measure* or *Poisson point measure,* if each variable $N(A)$ follows Poisson distribution $\mathcal{P}(\mu(A))$ and $N(\cdot)$ is an independently scattered measure. Recall that the definition of independently scattered measure assumes that for any disjoint sets $A_1, \ldots, A_m$ the random variables $N(A_1), \ldots, N(A_m)$ are independent and

$$N\left(\cup_{j=1}^{m} A_j\right) = \sum_{j=1}^{m} N(A_j) \qquad \text{almost surely.}$$

The measure μ is called *intensity measure* for N.

The family of random variables $\{\widetilde{N}(A) := N(A) - \mu(A), A \in \mathbf{A}\}$ is called *centered Poisson random measure.*

One can alternatively view Poisson random measure as a "locally finite" subset of $\mathcal{R}$, i.e. a random configuration of points (x_j) such that $N(A) = \#\{j : x_j \in A\}$. Here and elsewhere $\#V$ denotes the number of elements in a set V and "locally finite" means that the number of points in any set of finite measure is finite almost surely.

If intensity measure is finite, we can give an explicit construction of a Poisson random measure.

Proposition 7.1. *Assume that $\mu(\mathcal{R}) < \infty$. Let M be a Poisson random variable of intensity $\mu(\mathcal{R})$. Let (U_j) be a sequence (independent of M) of*

independent random elements of $\mathcal{R}$ with distribution $\frac{\mu(\cdot)}{\mu(\mathcal{R})}$.
   Then

$$N(A) := \sum_{j=1}^{M} \mathbf{1}_{\{U_j \in A\}}$$

is a Poisson random measure of intensity μ.

Proof.   For any fixed A our construction of the variable $N(A)$ is exactly the special case of (2.5) with Bernoulli random variables (i.e. the variables taking values 1 and 0 with probabilities $\frac{\mu(A)}{\mu(\mathcal{R})}$ and $1 - \frac{\mu(\cdot)}{\mu(\mathcal{R})}$, respectively). In this particular case (2.6) yields the characteristic function

$$\mathbb{E}\,e^{itN(A)} = \exp\left\{\mu(A)(e^{it} - 1)\right\}$$

corresponding to the Poisson distribution of intensity $\mu(A)$.

   It remains to prove the independence of $N(\cdot)$ for disjoint sets. Just for the sake of brevity, let us consider only two disjoint sets A_1, A_2. Let $n_1, n_2 \geq 0$. Then the event $\{N(A_1) = n_1, N(A_2) = n_2\}$ occurs whenever $M \geq n_1 + n_2$ and among the points $\{U_1, \ldots, U_M\}$ we have n_1 points in A_1, n_2 points in A_2, and $M - n_1 - n_2$ outside of both sets. The number of ways one can split the set of indices $\{1, \ldots, M\}$ into three groups of sizes $n_1, n_2, M - n_1 - n_2$ is equal to $\frac{M!}{n_1!n_2!(M-n_1-n_2)!}$. Once the splitting is chosen, the probability for the random elements with indices from the first group to belong to A_1, from the second to A_2, from the third to neither of the two, is equal to $p_1^{n_1} p_2^{n_2} (1 - p_1 - p_2)^{M-n_1-n_2}$, where we let $p_1 = \mathbb{P}(U_j \in A_1)$, $p_2 = \mathbb{P}(U_j \in A_2)$.

   Therefore,

$$\mathbb{P}\{N(A_1) = n_1, N(A_2) = n_2\}$$

$$= \sum_{m=n_1+n_2}^{\infty} \mathbb{P}(M = m) \cdot \frac{m!}{n_1!n_2!(m - n_1 - n_2)!} \cdot p_1^{n_1} p_2^{n_2} (1 - p_1 - p_2)^{m-n_1-n_2}$$

$$= e^{-\mu(\mathcal{R})} \cdot \frac{p_1^{n_1} p_2^{n_2}}{n_1!n_2!} \sum_{m=n_1+n_2}^{\infty} \frac{\mu(\mathcal{R})^m}{m!} \cdot \frac{m!}{(m - n_1 - n_2)!} \cdot (1 - p_1 - p_2)^{m-n_1-n_2}$$

$$= e^{-\mu(\mathcal{R})} \cdot \frac{p_1^{n_1} p_2^{n_2}}{n_1!n_2!} \cdot \mu(\mathcal{R})^{n_1+n_2} \sum_{\ell=0}^{\infty} \mu(\mathcal{R})^\ell \cdot \frac{(1 - p_1 - p_2)^\ell}{\ell!}$$

$$= e^{-\mu(\mathcal{R})} \cdot \frac{p_1^{n_1} p_2^{n_2}}{n_1!n_2!} \cdot \mu(\mathcal{R})^{n_1+n_2} \exp\left(\mu(\mathcal{R})(1 - p_1 - p_2)\right).$$

It follows that

$$\mathbb{P}\{N(A_1) = n_1, N(A_2) = n_2\}$$
$$= \frac{\mu(\mathcal{R})^{n_1} p_1^{n_1} \exp\left(-\mu(\mathcal{R})p_1\right)}{n_1!} \cdot \frac{\mu(\mathcal{R})^{n_2} p_2^{n_2} \exp\left(-\mu(\mathcal{R})p_2\right)}{n_2!}$$
$$= \frac{\mu(A_1)^{n_1} \exp\left(-\mu(A_1)\right)}{n_1!} \cdot \frac{\mu(A_2)^{n_2} \exp\left(-\mu(A_2)\right)}{n_2!}$$
$$= \mathbb{P}\{N(A_1) = n_1\} \cdot \mathbb{P}\{N(A_2) = n_2\},$$

as required for independence.   □

We will now build the it integrals of the functions $f : \mathcal{R} \mapsto \mathbb{R}$ with respect to the random measures N and $\widetilde{N}$. Let us first do it for step functions by letting, for $f := \sum_{j=1}^{m} c_j \mathbf{1}_{\{A_j\}}$ with $A_j \in \mathbf{A}$,

$$I_f := \int_{\mathcal{R}} f dN := \sum_{j=1}^{m} c_j N(A_j),$$

$$\widetilde{I}_f := \int_{\mathcal{R}} f d\widetilde{N} := \sum_{j=1}^{m} c_j \widetilde{N}(A_j).$$

Notice that in case of $\widetilde{N}(\cdot)$ there is nothing new in this definition. Indeed, since independent variables with finite variances are uncorrelated, $\widetilde{N}(\cdot)$ is a random measure with uncorrelated values. The corresponding intensity measure is μ, since by (1.13),

$$\mathbb{V}ar\,\widetilde{N}(A) = \mathbb{V}ar\,N(A) = \mu(A).$$

Therefore, the integral $\int_{\mathcal{R}} f d\widetilde{N}$ coincides so far with the one given by general construction described above in Subsection 7.1. We trivially have the formulae for expectations,

$$\mathbb{E} \int_{\mathcal{R}} f dN = \sum_{j=1}^{m} c_j \mathbb{E}\,N(A_j)$$

$$= \sum_{j=1}^{m} c_j \mu(A_j) = \int_{\mathcal{R}} f d\mu;$$

$$\mathbb{E} \int_{\mathcal{R}} f d\widetilde{N} = \sum_{j=1}^{m} c_j \mathbb{E}\,\widetilde{N}(A_j) = 0.$$

Moreover, it is clear that I_f and $\widetilde{I}_f$ differ just by centering,

$$\widetilde{I}_f = \sum_{j=1}^{m} c_j N(A_j) - \sum_{j=1}^{m} c_j \mu(A_j) = I_f - \int_{\mathcal{R}} f d\mu. \qquad (7.15)$$

Other way round,

$$I_f = \tilde{I}_f + \int_{\mathcal{R}} f \, d\mu. \tag{7.16}$$

We already know from Subsection 7.1 that the definition of the centered integral is correct: it does not depend of a particular representation of a step function as a sum. It is now clear from (7.16) that the same is true for the non-centered integral I_f.

Recall that by (7.6), (7.7) the centered integral is a linear functional, i.e.

$$\int_{\mathcal{R}} (cf) d\tilde{N} = c \int_{\mathcal{R}} f d\tilde{N},$$

$$\int_{\mathcal{R}} (f + g) d\tilde{N} = \int_{\mathcal{R}} f d\tilde{N} + \int_{\mathcal{R}} g d\tilde{N}.$$

Expression (7.16) immediately yields the similar property for non-centered integral.

The general theory also provides the formulae for variance, cf. (7.5),

$$\mathbb{V}ar \int_{\mathcal{R}} f d\tilde{N} = \int_{\mathcal{R}} f^2 d\mu$$

and covariance, cf. (7.8),

$$\mathrm{cov} \left(\int_{\mathcal{R}} f d\tilde{N}, \int_{\mathcal{R}} g d\tilde{N} \right) = \int_{\mathcal{R}} f g \, d\mu.$$

Again, by (7.16) the same formulae hold for non-centered Poisson integral.

Finally, we derive the expressions for characteristic functions. There is no loss of generality to assume that the step sets A_j are disjoint. Then, by using independence of $N(A_j)$ and the formula for Poisson characteristic function (1.16), we obtain

$$\mathbb{E} \exp \left(it \int_{\mathcal{R}} f dN \right) = \mathbb{E} \exp \left(it \sum_{j=1}^{m} c_j N(A_j) \right)$$

$$= \prod_{j=1}^{m} \mathbb{E} \exp \left(it c_j N(A_j) \right)$$

$$= \prod_{j=1}^{m} \exp \left((e^{it c_j} - 1) \mu(A_j) \right)$$

$$= \exp \left(\sum_{j=1}^{m} (e^{it c_j} - 1) \mu(A_j) \right)$$

$$= \exp \left(\int_{\mathcal{R}} (e^{itf} - 1) d\mu \right) \tag{7.17}$$

and by (7.15)

$$\mathbb{E} \exp\left(it \int_{\mathcal{R}} f d\widetilde{N}\right) = \mathbb{E} \exp\left(it\left(\int_{\mathcal{R}} f dN - \int_{\mathcal{R}} f d\mu\right)\right)$$

$$= \exp\left(\int_{\mathcal{R}} (e^{itf} - 1 - itf) d\mu\right). \qquad (7.18)$$

Now, once the integrals of step functions are well understood, we have to extend the notions of these integrals for general functions f satisfying some integrability condition. We will show that the non-centered integral extends to functions satisfying

$$\int_{\mathcal{R}} \min\{|f|, 1\} d\mu < \infty. \qquad (7.19)$$

For the centered integral, the general theory of Subsection 7.1 provides an extension to the integrands $f \in \mathbb{L}_2(\mathcal{R}, \mu)$. However, in Poissonian case, the integral will be well defined under weaker condition

$$\int_{\mathcal{R}} \min\{|f|^2, |f|\} d\mu < \infty. \qquad (7.20)$$

Both these extension domains can be understood by looking at the corresponding characteristic functions (7.17) and (7.18). Conditions (7.19), resp. (7.20) just describe the classes of functions for which the integrals in (7.17), resp. (7.18) are trivially finite.

It is useful to establish connections of the distributions of Poisson integrals with classes $\mathcal{P}_{1,0}$ and $\mathcal{P}_{2,1}$ introduced in Section 2. Let $\nu := \mu f^{-1}$ be a measure on $\mathbb{R}\backslash\{0\}$ defined by

$$\nu(B) := \mu f^{-1}(B), \qquad B \in \mathcal{B}.$$

Then by the variable change $u = f(r)$ for $r \in \mathcal{R}$, the formula for characteristic function (7.17) transforms into

$$\mathbb{E} \exp\left(it \int_{\mathcal{R}} f dN\right) = \exp\left(\int_{\mathbb{R}} (e^{itu} - 1)\nu(du)\right),$$

while the similar formula (7.18) becomes

$$\mathbb{E} \exp\left(it \int_{\mathcal{R}} f d\widetilde{N}\right) = \exp\left(\int_{\mathbb{R}} (e^{itu} - 1 - itu)\nu(du)\right). \qquad (7.21)$$

This corresponds exactly to (2.9), resp. to (2.11).

As for the integrability conditions, (7.19) becomes (2.8), while (7.20) becomes (2.10).

We conclude that the distribution of a non-centered Poisson integral belongs to the class $\mathcal{P}_{1,0}$, while the distribution of a centered Poisson integral belongs to the class $\mathcal{P}_{2,1}$.

We are going now to rigorously justify the extension of Poisson integrals from the class of step functions to the classes of integrands described in (7.19), resp. (7.20). A reader uninterested in technicalities may skip this part and just believe that those extensions do exist.

The key new point for extension is the definition of the non-centered Poisson integral of *arbitrary* measurable function $f : \mathcal{R} \mapsto \mathbb{R}$ under additional assumption $\mu(\mathcal{R}) < \infty$.

Therefore, assume for a while that $\mu(\mathcal{R}) < \infty$. Then any measurable function f admits an approximation by step functions f_n such that for any $\varepsilon > 0$

$$\lim_{n \to \infty} \mu\{u : |f(u) - f_n(u)| > \varepsilon\} = 0. \tag{7.22}$$

Indeed, we may take

$$f_n(u) = \sum_{j=-n^2}^{n^2} \frac{j}{n} \mathbf{1}_{\{\frac{j}{n} \leq f(u) < \frac{j+1}{n}\}}, \qquad u \in \mathcal{R}.$$

Then, since $\mu(\mathcal{R}) < \infty$,

$$\mu\left\{u : |f(u) - f_n(u)| > \tfrac{1}{n}\right\} \leq \mu\{u : |f(u)| > n\} \to 0, \qquad \text{as } n \to \infty,$$

and (7.22) clearly holds.

Let us fix an approximating sequence (f_n) satisfying (7.22) and show that the corresponding integrals

$$I_n := \int_{\mathcal{R}} f_n dN$$

converge in probability to some random variable that we, of course, designate to be $\int_{\mathcal{R}} f dN$. According to properties of Ky Fan distance (cf. Subsection 1.9), to justify this convergence, it is sufficient to show that (I_n) is a Cauchy sequence as defined in (1.20).

To this aim, the following simple estimate will be useful.

Lemma 7.2. *Let f be a step function. Then for any $\varepsilon, h > 0$ we have*

$$\mathbb{P}\left(\left|\int_{\mathcal{R}} f dN\right| > \varepsilon + \mu(\mathcal{R})h\right) \leq \mu\{u : |f(u)| > h\} + \frac{h^2 \mu(\mathcal{R})}{\varepsilon^2}. \tag{7.23}$$

Proof.   We have $f = f_1 + f_2$, where

$$f_1 := f\,\mathbf{1}_{\{|f|>h\}} := \sum_{j=1}^{m} c_j \mathbf{1}_{\{A_j\}}$$

with some disjoint step sets (A_j) and

$$f_2 := f\,\mathbf{1}_{\{|f|\le h\}} = f - f_1.$$

Then

$$I_1 := \int_{\mathcal{R}} f_1 dN = \sum_{j=1}^{m} c_j N(A_j)$$

and

$$\mathbb{P}(I_1 \neq 0) \le \sum_{j=1}^{m} \mathbb{P}(N(A_j) \neq 0)$$

$$= \sum_{j=1}^{m}(1 - e^{-\mu(A_j)}) \le \sum_{j=1}^{m} \mu(A_j) = \mu\{u : |f(u)| > h\}.$$

On the other hand, using the formulae for expectation and variance for Poisson integral

$$I_2 := \int_{\mathcal{R}} f_2 dN$$

we have

$$|\mathbb{E}\,I_2| = \left|\int_{\mathcal{R}} f_2 d\mu\right| \le \int_{\mathcal{R}} |f_2| d\mu \le \mu(\mathcal{R})h$$

and

$$\mathbb{V}ar I_2 = \int_{\mathcal{R}} f_2^2 d\mu \le \mu(\mathcal{R})h^2.$$

Finally, by using Chebyshev inequality (1.12), we get

$$\mathbb{P}\left(\left|\int_{\mathcal{R}} f dN\right| > \varepsilon + \mu(\mathcal{R})h\right) = \mathbb{P}\left(|I_1 + I_2| > \varepsilon + \mu(\mathcal{R})h\right)$$

$$\le \mathbb{P}(I_1 \neq 0) + \mathbb{P}\left(|I_2| > \varepsilon + \mu(\mathcal{R})h\right)$$

$$\le \mathbb{P}(I_1 \neq 0) + \mathbb{P}\left(|I_2 - \mathbb{E}\,I_2| > \varepsilon\right)$$

$$\le \mathbb{P}(I_1 \neq 0) + \frac{\mathbb{V}ar I_2}{\varepsilon^2}$$

$$\le \mu\{u : |f(u)| > h\} + \frac{h^2 \mu(\mathcal{R})}{\varepsilon^2},$$

as required.   □

Back to the approximating sequence of step functions $f_n \to f$, take any $\delta > 0$, fix any $\varepsilon < \delta$, then choose $h > 0$ so small that $\varepsilon + \mu(\mathcal{R})h \le \delta$ and $\frac{h^2 \mu(\mathcal{R})}{\varepsilon^2} \le \delta/2$. Then, using (7.22), choose n so large that for any $n_1 \ge n$

$$\mu\{u : |f(u) - f_{n_1}(u)| > h/2\} \le \delta/4.$$

Clearly, for any $n_1, n_2 \ge n$ we have

$$\mu\{u : |f_{n_1}(u) - f_{n_2}(u)| > h\} \le \delta/2.$$

By applying (7.23) to the difference $f_{n_1} - f_{n_2}$, we have

$$\mathbb{P}\left(|I_{n_1} - I_{n_2}| \ge \delta\right) \le \delta/2 + \delta/2 = \delta.$$

This means that $d_{KF}(I_{n_1}, I_{n_2}) \le \delta$, and proves that (I_n) is a Cauchy sequence with respect to Ky Fan distance. By completeness of the space of random variables w.r.t. that distance, we see that (I_n) converges in probability to a limiting random variable I.

Obviously, the limit I does not depend on the approximating sequence f_n. In fact, let (f'_n) and (f''_n) be two approximating sequences for f leading to the limits of integrals I', resp. I''. Then a sequence (f_n) defined by interlacement, $f_{2n-1} := f'_n$, $f_{2n} := f''_n$, also approximates f in the sense of (7.22) and leads to some limit of integrals I. Since $I = I'$, $I = I''$, we obtain $I' = I''$. Now, without any ambiguity we may let

$$\int_{\mathcal{R}} f dN := I.$$

Next, when defining this integral we may relax the assumption $\mu(\mathcal{R}) < \infty$ to

$$\mu\{u : f(u) \ne 0\} < \infty, \tag{7.24}$$

because the remaining part of the space, $\{u : f(u) = 0\}$ is not involved in the integral's definition at all.

Finally, we may define both Poisson integrals in full generality, as follows. If f obeys assumption (7.19), we let

$$\int_{\mathcal{R}} f dN := \int_{\mathcal{R}} f \mathbf{1}_{\{|f| \le 1\}} dN + \int_{\mathcal{R}} f \mathbf{1}_{\{|f| > 1\}} dN$$

$$:= \int_{\mathcal{R}} f \mathbf{1}_{\{|f| \le 1\}} d\tilde{N} + \int_{\mathcal{R}} f \mathbf{1}_{\{|f| \le 1\}} d\mu + \int_{\mathcal{R}} f \mathbf{1}_{\{|f| > 1\}} dN.$$

Here the first integral is well defined as an integral of a square integrable function over a measure with uncorrelated values; to justify this, recall that under (7.19)

$$\int_{\mathcal{R}} \left[f \mathbf{1}_{\{|f| \le 1\}}\right]^2 d\mu \le \int_{\mathcal{R}} |f| \mathbf{1}_{\{|f| \le 1\}} d\mu < \infty.$$

The second, deterministic, integral is well defined due to (7.19). Finally, the third integral is well defined because it fits into (7.24); indeed, under (7.19)

$$\mu\{u : f(u)\,\mathbf{1}_{\{|f(u)|>1\}} \neq 0\} = \mu\{u : |f(u)| > 1\} < \infty.$$

Assume now that f obeys assumption (7.20). In this case, we let

$$\int_{\mathcal{R}} f d\widetilde{N} := \int_{\mathcal{R}} f\,\mathbf{1}_{\{|f|\leq 1\}} d\widetilde{N} + \int_{\mathcal{R}} f\,\mathbf{1}_{\{|f|>1\}} d\widetilde{N}$$

$$:= \int_{\mathcal{R}} f\,\mathbf{1}_{\{|f|\leq 1\}} d\widetilde{N} + \int_{\mathcal{R}} f\,\mathbf{1}_{\{|f|>1\}} dN - \int_{\mathcal{R}} f\,\mathbf{1}_{\{|f|>1\}} d\mu.$$

Here the first integral is well defined as an integral of a square integrable function over a measure with uncorrelated values because under (7.20)

$$\int_{\mathcal{R}} \left[f\,\mathbf{1}_{\{|f|\leq 1\}} \right]^2 d\mu = \int_{\mathcal{R}} |f|^2 \,\mathbf{1}_{\{|f|\leq 1\}} d\mu < \infty.$$

The second integral is well defined because it fits into (7.24); indeed, under (7.20),

$$\mu\{u : f(u)\,\mathbf{1}_{\{|f(u)|>1\}} \neq 0\} = \mu\{u : |f(u)| > 1\} < \infty.$$

The third, deterministic, integral is well defined due to (7.20) either.

It is a routine exercise to check that the properties that we have proved for the integrals of step functions remain valid for extended classes of integrals, of course, under the corresponding natural assumptions. In particular, if the integrand f obeys (7.19), then the formula for characteristic function (7.17) holds true. We also have

$$\mathbb{E} \int_{\mathcal{R}} f dN = \int_{\mathcal{R}} f d\mu$$

whenever the integral on the right hand side is well defined, i.e. it is true that $f \in \mathrm{L}_1(\mathcal{R}, \mu)$, and

$$\mathrm{Var} \int_{\mathcal{R}} f dN = \int_{\mathcal{R}} f^2 d\mu \qquad (7.25)$$

whenever the integral on the right hand side is well defined, i.e. it is true that $f \in \mathrm{L}_2(\mathcal{R}, \mu)$.

If the integrand f obeys (7.20), then the formula for characteristic function (7.18) holds true. We also have

$$\mathbb{E} \int_{\mathcal{R}} f d\widetilde{N} = 0$$

for any function of this class and

$$\mathbb{V}ar \int_{\mathcal{R}} f d\widetilde{N} = \int_{\mathcal{R}} f^2 d\mu$$

whenever the integral on the right hand side is well defined, i.e. it is true that $f \in \mathbb{L}_2(\mathcal{R}, \mu)$.

Moreover, for covariances we have

$$\text{cov}\left(\int_{\mathcal{R}} f dN, \int_{\mathcal{R}} g dN\right) = \text{cov}\left(\int_{\mathcal{R}} f d\widetilde{N}, \int_{\mathcal{R}} g d\widetilde{N}\right) = \int_{\mathcal{R}} f g d\mu \quad (7.26)$$

whenever $f, g \in \mathbb{L}_2(\mathcal{R}, \mu)$.

Exercise 7.3. Let $\mathcal{R} := \mathbb{R}_+ \times \mathbb{R}^d$ and let N be a Poisson random measure of intensity $\mu := \frac{dr}{r^{1+\alpha}} \nu(du)$ with some measure ν on $\mathbb{R}^d$. Assume that a random process $X(t), t \in T$ admits an integral representation

$$X(t) = \int_{\mathcal{R}} r f_t(u) \, dN, \qquad t \in T.$$

Prove that X is an α-stable process.

7.5 *Independently scattered stable random measures and integrals*

The following construction reminds very much that of the Gaussian white noise and of the corresponding integral. We only replace the centered normal random variables with strictly α-stable ones.

As we defined in Subsection 2.5, a random variable X (and its distribution) is strictly α-stable, if for any independent variables X_1 and X_2 equidistributed with X and for any $k_1, k_2 > 0$ the variable $k_1 X_1 + k_2 X_2$ is equidistributed with $k_3 X$ where

$$k_3 = (k_1^\alpha + k_2^\alpha)^{1/\alpha}. \quad (7.27)$$

Moreover, according to (2.24) and (2.20), the characteristic function of X with parameters α, c_-, c_+ can be written in the form

$$\mathbb{E} \, e^{itX} = \exp\{\Psi(t)\},$$

where

$$\Psi(t) = \begin{cases} \int (e^{itu} - 1) \frac{c(u)du}{|u|^{\alpha+1}}, & 0 < \alpha < 1, \\ iat - c|t|, & \alpha = 1, \\ \int (e^{itu} - 1 - itu) \frac{c(u)du}{|u|^{\alpha+1}}, & 1 < \alpha < 2, \end{cases} \quad (7.28)$$

and

$$c(u) = \begin{cases} c_-, & u < 0, \\ c_+, & u > 0. \end{cases}$$

We will denote $\widetilde{\mathcal{S}}(c_-, c_+, \alpha)$ and $\widetilde{\mathcal{S}}(a, c, 1)$, the distribution of the random variable with log-characteristic function (7.28) in cases $\alpha \neq 1$ and $\alpha = 1$, respectively.

In the rest of the section, just for notation uniformity, we handle the case $\alpha \neq 1$ but the results and arguments exposed below remain valid for $\alpha = 1$, too.

Recall the scaling and summation rules: for any $k > 0$ we have

$$P_{kX} = \widetilde{\mathcal{S}}(k^\alpha c_-, k^\alpha c_+, \alpha) \tag{7.29}$$

and

$$P_{-kX} = \widetilde{\mathcal{S}}(k^\alpha c_+, k^\alpha c_-, \alpha). \tag{7.30}$$

If X and X' are strictly stable variables with distributions $\widetilde{\mathcal{S}}(c_+, c_-, \alpha)$, resp. $\widetilde{\mathcal{S}}(c_+', c_-', \alpha)$, then $X + X'$ is also strictly stable and

$$P_{X+X'} = \widetilde{\mathcal{S}}(c_+ + c_+', c_- + c_-', \alpha). \tag{7.31}$$

This is obvious from the formulae for log-characteristic functions $\Psi(\cdot)$ given in (7.28). Notice that this is also a particular case of summation rule for triplets (cf. the end of Subsection 2.4) specified to stable distributions.

For this subsection, we fix parameters α, c_-, c_+.

As usual, let $(\mathcal{R}, \mu)$ be a measure space and $\mathbf{A} = \{A \subset \mathcal{R}, \mu(A) < \infty\}$. A family of random variables $\{X(A), A \in \mathbf{A}\}$ is called *independently scattered α-stable random measure* with parameters c_-, c_+, if each variable $X(A)$ follows distribution $\widetilde{\mathcal{S}}(\mu(A)c_-, \mu(A)c_+, \alpha)$ and $X(\cdot)$ is an independently scattered measure. Recall that the definition of independently scattered measure assumes that for any disjoint sets $A_1, \ldots, A_m$ the random variables $X(A_1), \ldots, X(A_m)$ are independent and

$$X\left(\cup_{j=1}^m A_j\right) = \sum_{j=1}^m X(A_j) \qquad \text{almost surely.} \tag{7.32}$$

The measure μ is called *intensity measure* for X.

Notice that summation property (7.32) is in full agreement with the scaling and strict stability. It means that by scaling rule (7.29) $X(A_j)$ is equidistributed with $\mu(A_j)^{1/\alpha}\tilde{X}$, where $\tilde{X}$ is a random variable with distribution $\widetilde{S}(c_-, c_+, \alpha)$. By (7.27) we obtain that the right hand side of (7.32) is equidistributed with

$$\left(\sum_{j=1}^m \mu(A_j) \right)^{1/\alpha} \tilde{X} = \left(\mu \left(\cup_{j=1}^m A_j \right) \right)^{1/\alpha} \tilde{X},$$

exactly as, by scaling, the left hand side of (7.32).

Let us define the *integrals* for functions $f : \mathcal{R} \mapsto \mathbb{R}$ with respect to X. For a step function $f := \sum_{j=1}^m c_j \mathbf{1}_{\{A_j\}}$ the integral is defined in a natural way, as before:

$$\int_{\mathcal{R}} f dX := \sum_{j=1}^m c_j X(A_j). \tag{7.33}$$

Due to the additivity property, the definition of the integral is correct: it does not depend of a particular representation of a step function as a sum.

The constructed integral is a linear functional, i.e.

$$\int_{\mathcal{R}} (cf) dX = c \int_{\mathcal{R}} f dX,$$

$$\int_{\mathcal{R}} (f + g) dX = \int_{\mathcal{R}} f dX + \int_{\mathcal{R}} g dX.$$

Let us compute the distribution of the integral (7.33). First, we do it for a *non-negative* function f. There is no loss of generality in assuming that the sets A_j are disjoint. By using the scaling rule (7.29), we see that $\mu(A_j)^{1/\alpha} \tilde{X}$ has the same distribution $\widetilde{S}(\mu(A)c_-, \mu(A)c_+, \alpha)$ as $X(A_j)$. Hence, $c_j X(A_j)$ is equidistributed with $c_j \mu(A_j)^{1/\alpha} \tilde{X}$. Next, using independence of $X(A_j)$ and (7.27) we see that the sum in (7.33) is equidistributed with

$$\left(\sum_{j=1}^m c_j^\alpha \mu(A_j) \right)^{1/\alpha} \tilde{X} = \|f\|_\alpha \tilde{X},$$

where we recall the classical notation

$$\|f\|_\alpha := \left(\int_{\mathcal{R}} |f|^\alpha d\mu \right)^{1/\alpha}.$$

Finally, using again the scaling rule (7.29) we infer that the distribution of the integral (7.33) for a non-negative function f is $\widetilde{S}(\|f\|_\alpha^\alpha c_-, \|f\|_\alpha^\alpha c_+, \alpha)$.

If a function f is non-positive, we take into account the flipping effect (7.30), where c_- and c_+ exchange their places, and obtain that the distribution of the integral (7.33) for such function is $\widetilde{S}(\|f\|_\alpha^\alpha c_+, \|f\|_\alpha^\alpha c_-, \alpha)$.

As for an arbitrary step function f, we may split it into a difference of the two positive functions with disjoint supports, $f = f_+ - f_-$, where

$$f_+(u) := \max\{f(u), 0\},$$
$$f_-(u) := \max\{-f(u), 0\}.$$

We write

$$\int_{\mathcal{R}} f dX = \int_{\mathcal{R}} f_+ dX + \int_{\mathcal{R}} (-f_-) dX,$$

and notice that the integrals on the right hand side are independent because the integrands have disjoint supports on which the values of X are independent. By using the rule (7.31) we conclude that the distribution of the integral (7.33) in general case is equal to

$$\widetilde{\mathcal{S}}(\|f_+\|_\alpha^\alpha c_- + \|f_-\|_\alpha^\alpha c_+, \|f_+\|_\alpha^\alpha c_+ + \|f_-\|_\alpha^\alpha c_-, \alpha). \qquad (7.34)$$

For symmetric case, $c_- = c_+ := c$, we may use the identity

$$\|f_+\|_\alpha^\alpha + \|f_-\|_\alpha^\alpha = \|f\|_\alpha^\alpha$$

and obtain a nicer distribution

$$\widetilde{\mathcal{S}}(\|f\|_\alpha^\alpha c, \|f\|_\alpha^\alpha c, \alpha).$$

Therefore, by the scaling rule in symmetric case the integral (7.33) is equidistributed with $\|f\|_\alpha \widetilde{X}$ for any step function f.

By using approximation by step functions it is not hard to define the integral

$$\int_{\mathcal{R}} f dX$$

for all functions $f \in \mathbb{L}_\alpha(\mathcal{R}, \mu)$, i.e. for the functions satisfying

$$\|f\|_\alpha^\alpha = \int_{\mathcal{R}} |f|^\alpha d\mu < \infty.$$

The linearity and the formula for distribution (7.34) are still true for this larger class of integrands.

Exercise 7.4. Let $f_1, \ldots, f_n \in \mathbb{L}_\alpha(\mathcal{R}, \mu)$, and let X be an independently scattered α-stable random measure on $(\mathcal{R}, \mu)$. Prove that $\left(\int_{\mathcal{R}} f_j dX \right)_{1 \leq j \leq n}$ is an α-stable random vector according to the definition given in Section 5.

Exercise 7.5. Check that the construction of integral works in the same way for $\alpha = 1$.

The technical details of integral's extension are given below. A reader uninterested in technicalities may skip this part.

The extension of the integral to the integrands from $\mathbb{L}_\alpha(\mathcal{R}, \mu)$ goes along the same lines as that of Poisson integrals. Any function $f \in \mathbb{L}_\alpha(\mathcal{R}, \mu)$ admits an approximation by step functions f_n such that

$$\lim_{n \to \infty} ||f - f_n||_\alpha = 0. \tag{7.35}$$

For any approximating sequence (f_n) satisfying (7.35) one shows that the corresponding integrals

$$I_n := \int_\mathcal{R} f_n dN$$

converge in probability to some random variable I that is designated to be $\int_\mathcal{R} f dN$. According to properties of Ky Fan distance d_{KF} (cf. Subsection 1.9), to justify this convergence, it is sufficient to show that (I_n) is a Cauchy sequence as defined in (1.20).

The proof of Cauchy property is based on the following result replacing Lemma 7.2.

Lemma 7.6. *Let X be an independently scattered α-stable random measure with parameters c_-, c_+. Let f be a step function. Then for any $\varepsilon > 0$ we have*

$$\mathbb{P}\left(\left|\int_\mathcal{R} f dX\right| > \varepsilon\right) \le 2\,\mathbb{P}\left(|Y| > \frac{\varepsilon}{2||f||_\alpha(c_- + c_+)^{1/\alpha}}\right), \tag{7.36}$$

where Y is a strictly stable variable with distribution $\widetilde{\mathcal{S}}(0, 1, \alpha)$.

Proof. We know that $\int_\mathcal{R} f dX$ has the distribution (7.34) which by (7.31) and (7.29) coincides with the distribution of

$$(||f_+||_\alpha^\alpha c_+ + ||f_-||_\alpha^\alpha c_-)^{1/\alpha} Y_1 - (||f_+||_\alpha^\alpha c_- + ||f_-||_\alpha^\alpha c_+)^{1/\alpha} Y_2,$$

where Y_1, Y_2 are independent copies of Y. Hence,

$$\mathbb{P}\left(\left|\int_\mathcal{R} f dN\right| > \varepsilon\right) \le \mathbb{P}\left((||f_+||_\alpha^\alpha c_+ + ||f_-||_\alpha^\alpha c_-)^{1/\alpha} |Y_1| > \varepsilon/2\right)$$

$$+ \mathbb{P}\left((||f_+||_\alpha^\alpha c_- + ||f_-||_\alpha^\alpha c_+)^{1/\alpha} |Y_2| > \varepsilon/2\right)$$

$$\le 2\,\mathbb{P}\left(||f||_\alpha(c_- + c_+)^{1/\alpha}|Y| > \varepsilon/2\right)$$

$$= 2\,\mathbb{P}\left(|Y| > \frac{\varepsilon}{2||f||_\alpha(c_- + c_+)^{1/\alpha}}\right),$$

as required. □

Given a $\delta > 0$ we obtain from (7.36) that for any n_1, n_2 large enough (fill in a few missing details!)

$$\mathbb{P}\left(|I_{n_1} - I_{n_2}| \geq \delta\right) < \delta.$$

This means that $d_{KF}(I_{n_1}, I_{n_2}) \leq \delta$, and proves that (I_n) is a Cauchy sequence with respect to Ky Fan distance. By completeness of the space of random variables w.r.t. that distance, we see that (I_n) converges in probability to a limiting random variable I. By the same interlacement argument used in Poisson case one shows that the limit I does not depend of the approximating sequence f_n. Hence there is no ambiguity in letting

$$\int_{\mathcal{R}} f dN := I.$$

8   Limit Theorems for Poisson Integrals

8.1   *Convergence to the normal distribution*

Our aim here is to establish the conditions under which a sequence of centered Poisson integrals $\int_{\mathcal{R}} f_n d\widetilde{N}$ converges in distribution to a centered normal distribution $\mathcal{N}(0, \sigma^2)$. We start with characteristic functions. Recall that by (7.18)

$$\mathbb{E} \exp\left(it \int_{\mathcal{R}} f d\widetilde{N}\right) = \exp\left(\int_{\mathcal{R}} (e^{itf} - 1 - itf) d\mu\right)$$

for any f satisfying the key condition (7.20). Therefore, in the language of characteristic functions the convergence to the normal distribution takes the form

$$\int_{\mathcal{R}} (e^{itf_n} - 1 - itf_n) d\mu \to -\frac{\sigma^2 t^2}{2}, \qquad \forall t \in \mathbb{R}. \tag{8.1}$$

Proposition 8.1. *Let (f_n) be a sequence of functions from $\mathbb{L}_2(\mathcal{R}, \mu)$. Assume that for some $\sigma \geq 0$ it is true that*

$$\mathbb{V}ar \int_{\mathcal{R}} f_n dN = \int_{\mathcal{R}} f_n^2 d\mu \to \sigma^2 \tag{8.2}$$

and that for any $\varepsilon > 0$ it is true that

$$\int_{|f_n| > \varepsilon} f_n^2 d\mu \to 0. \tag{8.3}$$

Then the variables $\int_{\mathcal{R}} f_n d\widetilde{N}$ converge in distribution to the normal distribution $\mathcal{N}(0, \sigma^2)$.

Proof.   The key idea is to notice that for small u we have

$$e^{iu} - 1 - iu \approx (iu)^2/2 = -u^2/2.$$

More formally, for $|u| \leq 1$, we have

$$\left| e^{iu} - 1 - iu - \frac{i^2 u^2}{2} \right| \leq c|u|^3.$$

Let us fix an $\varepsilon > 0$. If $|t|\varepsilon \leq 1$, then we write

$$\int_{|f_n| \leq \varepsilon} \left(e^{it f_n} - 1 - it f_n \right) d\mu$$

$$= \int_{|f_n| \leq \varepsilon} \left(e^{it f_n} - 1 - it f_n - \frac{i^2 t^2 f_n^2}{2} \right) d\mu - \int_{|f_n| \leq \varepsilon} \frac{t^2 f_n^2}{2} d\mu.$$

The first term is small:

$$\int_{|f_n| \leq \varepsilon} \left| e^{it f_n} - 1 - it f_n - \frac{i^2 t^2 f_n^2}{2} \right| d\mu \leq \int_{|f_n| \leq \varepsilon} c|t|^3 |f_n|^3 d\mu$$

$$\leq c|t|^3 \varepsilon \int_{\mathcal{R}} f_n^2 d\mu \to c|t|^3 \sigma^2 \varepsilon.$$

The second term is convergent by (8.2) and (8.3) :

$$\int_{|f_n| \leq \varepsilon} \frac{t^2 f_n^2}{2} d\mu = \frac{t^2}{2} \left(\int_{\mathcal{R}} f_n^2 d\mu - \int_{|f_n| > \varepsilon} f_n^2 d\mu \right) \to \frac{\sigma^2 t^2}{2} \ .$$

On the other hand,

$$\int_{|f_n| \geq \varepsilon} \left| e^{it f_n} - it f_n - 1 \right| d\mu \leq \int_{|f_n| \geq \varepsilon} \left(2 + |t f_n| \right) d\mu$$

$$\leq \int_{|f_n| \geq \varepsilon} \left(2 \frac{f_n^2}{\varepsilon^2} + |t f_n| \frac{|f_n|}{\varepsilon} \right) d\mu$$

$$= \left(\frac{2}{\varepsilon^2} + \frac{|t|}{\varepsilon} \right) \int_{|f_n| \geq \varepsilon} f_n^2 \, d\mu \to 0.$$

By summing up, we have

$$\limsup_{n \to \infty} \left| \int_{\mathcal{R}} (e^{it f_n} - 1 - it f_n) d\mu + \frac{\sigma^2 t^2}{2} \right| \leq c|t|^3 \sigma^2 \varepsilon.$$

Finally, letting $\varepsilon \to 0$, we arrive at (8.1).   □

Notice that integrals are the simplest (linear) functionals of Poisson random measure. We refer to [78] for a much deeper theory of normal approximation of non-linear Poisson functionals.

8.2   *Convergence to a stable distribution*

By comparing characteristic functions (7.21) and (7.28), we see that an integral with respect to a centered Poisson random measure $\int f d\widetilde{N}$ has exactly the stable distribution $\widetilde{\mathcal{S}}(c_-, c_+, \alpha)$, if the integrand f has the corresponding distribution with respect to the intensity measure μ, that is for any Borel set $B \in \mathcal{B}$ we have

$$\mu\{f \in B\} := \mu\{r \in \mathcal{R} : f(r) \in B\} = \int_B \frac{c(u)du}{|u|^{\alpha+1}} \, ,$$

where as usual

$$c(u) = \begin{cases} c_-, & u < 0, \\ c_+, & u > 0. \end{cases}$$

These identities hold for any $B \in \mathcal{B}$ if and only if they are true for half-lines $B = [x, +\infty)$ and $B = (-\infty, -x]$ for all $x > 0$, i.e. iff

$$\mu\{f \geq x\} = \int_x^\infty \frac{c_+ du}{|u|^{\alpha+1}} = \frac{c_+}{\alpha x^\alpha} \, ,$$

$$\mu\{f \leq -x\} = \frac{c_-}{\alpha x^\alpha} \, .$$

Therefore, the main conditions providing the convergence in distribution of a sequence $\int f_n d\widetilde{N}$ to the distribution $\widetilde{\mathcal{S}}(c_-, c_+, \alpha)$, are

$$\lim_{n\to\infty} \mu\{f_n \geq x\} = \frac{c_+}{\alpha x^\alpha} \, , \qquad \forall x > 0, \tag{8.4}$$

$$\lim_{n\to\infty} \mu\{f_n \leq -x\} = \frac{c_-}{\alpha x^\alpha} \, , \qquad \forall x > 0. \tag{8.5}$$

These conditions, however, are not sufficient, since some problems might show up at zero and at infinity. We show what can happen with simplest examples where $c_- = c_+ = 0$.

Example 8.2. Let $f_n = n^{-1/2} \mathbf{1}_{\{A_n\}}$ and $\mu\{A_n\} = n$. Then the limits above are equal to zero but the integrals converge in distribution to the standard normal distribution according to Proposition 8.1.

Example 8.3. Let $f_n = n \mathbf{1}_{\{A_n\}}$ and $\mu\{A_n\} = n^{-1}$. The limits above are again zero but the integrals converge in distribution to -1, since the corresponding non-centered integrals converge to zero in distribution and their expectations are equal to one.

By taking these examples into account, one will not consider the following proposition as too cumbersome.

Proposition 8.4. *Let* $\alpha \in (1,2)$. *Assume that the relations* (8.4) *and* (8.5) *hold, as well as*

$$\lim_{\varepsilon \to 0} \limsup_{n \to \infty} \int_{|f_n| \leq \varepsilon} f_n^2 \, d\mu = 0 \tag{8.6}$$

and for any $\varepsilon > 0$ *it is true that*

$$\lim_{n \to \infty} \int_{|f_n| > \varepsilon} f_n d\mu = \int_{|u| > \varepsilon} \frac{u \, c(u) \, du}{|u|^{\alpha+1}} = \frac{c_+ - c_-}{(\alpha - 1)\varepsilon^{\alpha-1}} \cdot \tag{8.7}$$

Then the distributions of random variables $\int_{\mathcal{R}} f_n d\widetilde{N}$ *converge to the strictly stable distribution* $\widetilde{S}(c_-, c_+, \alpha)$.

Proof. We check the convergence of characteristic functions. Take a small $\varepsilon > 0$ and split the expression in the exponent (7.18) into three parts

$$\int_{|f_n| \leq \varepsilon} (e^{itf_n} - 1 - itf_n) d\mu + \int_{|f_n| > \varepsilon} (e^{itf_n} - 1) d\mu - it \int_{|f_n| > \varepsilon} f_n d\mu.$$

The third part converges to the corresponding component of (7.28) by (8.7).

The same is true for the second part. To show this, make a variable change $u = f_n(r)$ and integrate by parts. We have

As for the first part, by using the inequality

$$\left| e^{iu} - 1 - iu \right| \leq |u|^2/2$$

by choosing ε to be small enough and taking (8.6) into account, we can render it arbitrarily small both for our expressions and for the limit. $\quad\Box$

We can give a simpler sufficient condition for convergence.

Corollary 8.5. *Let* $\alpha \in (1,2)$. *Assume that relations* (8.4) *and* (8.5) *hold true and that for some* $n_0 \in \mathbb{N}, C > 0$ *we have a uniform bound*

$$\mu\{|f_n| \geq u\} \leq \frac{C}{u^\alpha}, \qquad \forall u > 0, \, \forall n \geq n_0. \tag{8.8}$$

Then the distributions of the variables $\int_{\mathcal{R}} f_n d\widetilde{N}$ *converge to the stable distribution* $\widetilde{S}(c_-, c_+, \alpha)$.

Proof.   Let us verify the assumptions of Proposition 8.4. By integrating by parts and using the uniform bound (8.8) we find for $n \geq n_0$

$$\int_{|f_n| \leq \varepsilon} f_n^2 d\mu = -\int_0^\varepsilon u^2 d\mu \{u \leq |f_n| \leq \varepsilon\}$$

$$= \varepsilon^2 \, \mu \{|f_n| = \varepsilon\} + 2\int_0^\varepsilon u\mu\{u \leq |f_n| \leq \varepsilon\} du$$

$$\leq \varepsilon^2 \, \mu \{|f_n| \geq \varepsilon\} + 2\int_0^\varepsilon u\mu\{u \leq |f_n|\} du$$

$$\leq C\varepsilon^{2-\alpha} + 2C\int_0^\varepsilon u^{1-\alpha} du = \frac{(4-\alpha)C\varepsilon^{2-\alpha}}{2-\alpha},$$

and (8.6) follows. In order to check (8.7), we integrate by parts again

$$\int_{f_n > \varepsilon} f_n d\mu = -\int_\varepsilon^\infty u \, d\mu\{f_n \geq u\} = \varepsilon\mu\{f_n \geq \varepsilon\} + \int_\varepsilon^\infty \mu\{f_n \geq u\} du.$$

Notice that by (8.8) we have

$$\lim_{u \to +\infty} u \, \mu\{f_n \geq u\} = 0,$$

thus another external term in the integration by part formula vanishes.

Now we let n go to infinity. Notice that the right hand side of (8.8) is summable on $[\varepsilon, +\infty)$, whenever $\alpha > 1$. Hence we may apply Lebesgue's majorated convergence theorem for integrals. By using the limit relation (8.4) we obtain

$$\lim_{n\to\infty} \int_{f_n > \varepsilon} f_n d\mu = \varepsilon \lim_{n\to\infty} \mu\{f_n \geq \varepsilon\} + \int_\varepsilon^\infty \lim_{n\to\infty} \mu\{f_n \geq u\} du$$

$$= \varepsilon \cdot \frac{c_+}{\alpha\,\varepsilon^\alpha} + \int_\varepsilon^\infty \frac{c_+\,du}{\alpha\,u^\alpha}$$

$$= \frac{c_+}{\alpha}\left(\varepsilon^{1-\alpha} + \frac{\varepsilon^{1-\alpha}}{\alpha-1}\right)$$

$$= \frac{c_+\,\varepsilon^{1-\alpha}}{\alpha}\left(1 + \frac{1}{\alpha-1}\right) = \frac{c_+\,\varepsilon^{1-\alpha}}{\alpha-1}.$$

On the negative half-line we obtain in the same way

$$\lim_{n\to\infty} \int_{f_n < -\varepsilon} f_n \, d\mu = \frac{-c_-\,\varepsilon^{1-\alpha}}{\alpha-1}.$$

By summing up the two limits we arrive at (8.7), as required. $\square$

We will also need the following "degenerated" version of Corollary 8.5.

Exercise 8.6. Let $\alpha \in (1,2)$. Assume that for some positive sequence (c_n) tending to zero we have

$$\mu\{|f_n| \geq u\} \leq \frac{c_n}{u^\alpha}, \qquad \forall u > 0.$$

Then $\int_\mathcal{R} f_n d\widetilde{N} \Rightarrow 0$.

9 Lévy Processes

9.1 *General Lévy processes*

A random process $Y(t), t \geq 0$, starting at zero and having stationary and independent increments is called *Lévy process*. Wiener and Poisson processes are the first obvious and fundamental examples of this class. One may intuitively perceive a general Lévy process as a sum of three parts: a linear function, a scaled Wiener process, and a "partially centered mixture of scaled Poisson processes".

There is a one to one correspondence between Lévy processes and infinitely divisible distributions discussed in Subsection 2.4. Indeed, let Y be a Lévy processes. Then for any $t \geq 0, n \in \mathbb{N}$, we have, using $Y(0) = 0$,

$$Y(t) = \sum_{j=1}^{n}(Y(\tfrac{jt}{n}) - Y(\tfrac{(j-1)t}{n})). \tag{9.1}$$

Notice that by definition of Lévy process the summands are independent and identically distributed. Hence, the variable $Y(t)$ is infinitely divisible. Let (a, σ^2, ν) be the triplet corresponding to the infinitely divisible distribution of $Y(1)$. Then by letting $t = 1$ in (9.1) and using summation rule for triplets (see Subsection 2.4) we conclude that the triplet corresponding to the distribution of $Y(\tfrac{1}{n})$ is $(\tfrac{a}{n}, \tfrac{\sigma^2}{n}, \tfrac{\nu}{n})$. Furthermore, for any $m \in \mathbb{N}$ by letting $t = \tfrac{m}{n}$ in (9.1) we see that the triplet corresponding to the distribution of $Y(\tfrac{m}{n})$ is $(\tfrac{am}{n}, \tfrac{\sigma^2 m}{n}, \tfrac{\nu m}{n}) = (ta, t\sigma^2, t\nu)$.

From now on we will assume that our Lévy processes are continuous in probability. Then by approximating arbitrary $t \geq 0$ with rational numbers we conclude that the triplet corresponding to the distribution of $Y(t)$ is also equal to $(ta, t\sigma^2, t\nu)$. For characteristic functions we obtain, according to (2.15), the expression

$$f_{Y(t)}(\tau) = \exp\left\{ t\left(ia\tau - \frac{\sigma^2 \tau^2}{2} + \int (e^{i\tau u} - 1 - i\tau u\, \mathbf{1}_{\{|u| \leq 1\}}) \nu(du) \right) \right\}$$

valid for all $t \geq 0, \tau \in \mathbb{R}$.

Furthermore, this formula determines all finite-dimensional distributions of Y, since for any times $0 = t_0 < t_1 < \cdots < t_n$ and for any $v \in \mathbb{R}^n$ we may easily calculate the characteristic function of the random vector

$Z := (Y(t_1), \ldots, Y(t_n))$ as follows. First, write

$$(v, Z) = \sum_{j=1}^{n} v_j Y(t_j) = \sum_{j=1}^{n} v_j \sum_{k=1}^{j} (Y(t_k) - Y(t_{k-1}))$$

$$= \sum_{k=1}^{n} \left[(Y(t_k) - Y(t_{k-1})) \sum_{j=k}^{n} v_j \right]$$

$$:= \sum_{k=1}^{n} [(Y(t_k) - Y(t_{k-1}))V_k],$$

then

$$f_Z(v) = \mathbb{E}\, e^{i(v, Z)} = \mathbb{E} \exp\left(i \sum_{k=1}^{n} [V_k(Y(t_k) - Y(t_{k-1}))] \right)$$

$$= \prod_{k=1}^{n} \mathbb{E} \exp\left(i\, [V_k(Y(t_k) - Y(t_{k-1}))] \right)$$

$$= \prod_{k=1}^{n} \mathbb{E} \exp\left(i V_k\, Y(t_k - t_{k-1}) \right)$$

$$= \exp\left(\sum_{k=1}^{n} (t_k - t_{k-1}) L(V_k) \right),$$

where

$$L(\tau) := ia\tau - \frac{\sigma^2 \tau^2}{2} + \int \left(e^{i\tau u} - 1 - i\tau u\, \mathbf{1}_{\{|u| \le 1\}} \right) \nu(du).$$

Summarizing, the triplet (a, σ^2, ν) of $Y(1)$ determines all finite-dimensional distributions of the process Y.

We show now how to construct a Lévy process corresponding to a given infinitely divisible distribution. Recall that by Lévy–Khinchin representation (2.15) any infinitely divisible distribution is characterized by a triplet (a, σ^2, ν) where $a \in \mathbb{R}$, $\sigma^2 \ge 0$, and a measure ν belongs to the class $\mathcal{P}_{2,0}$. Therefore, we have to build a Lévy process Y such that

$$\mathbb{E}\, e^{i\tau Y(1)} = \exp\left\{ ia\tau - \frac{\sigma^2 \tau^2}{2} + \int \left(e^{i\tau u} - 1 - i\tau u\, \mathbf{1}_{\{|u| \le 1\}} \right) \nu(du) \right\}. \quad (9.2)$$

The first two terms clearly correspond to a shifted and scaled Wiener process $at + \sigma W(t)$, cf. Subsection 6.1. Therefore, the only problem is to find a process corresponding to the integral part. Let $\mathcal{R} := \mathbb{R} \times \mathbb{R}_+$ and define a product type measure μ on $\mathcal{R}$ as $\mu(du, ds) := \nu(du)ds$. Let N be a Poisson random measure on $\mathbb{R}$ with intensity μ. We denote $\widetilde{N}$ the corresponding centered Poisson random measure. Let

$$Y_\nu(t) := \int_{\mathcal{R}} \mathbf{1}_{\{s \leq t\}} \mathbf{1}_{\{|u| > 1\}} \, u \, N(du, ds)$$

$$+ \int_{\mathcal{R}} \mathbf{1}_{\{s \leq t\}} \mathbf{1}_{\{|u| \leq 1\}} \, u \, \widetilde{N}(du, ds). \qquad (9.3)$$

Both integrals are correctly defined since $\nu \in \mathcal{P}_{2,0}$.

Exercise 9.1. Prove that conditions (7.19) and (7.20) providing the existence of Poisson integrals are verified for (9.3).

By the formulae for characteristic functions (7.17) and (7.18) we obtain

$$\mathbb{E} \, e^{i\tau Y_\nu(t)} = \exp \left\{ \int_0^t \int_{|u| > 1} (e^{i\tau u} - 1)\nu(du)ds \right.$$

$$\left. + \int_0^t \int_{|u| \leq 1} (e^{i\tau u} - 1 - i\tau u)\nu(du)ds \right\}$$

$$= \exp \left\{ t \int (e^{i\tau u} - 1 - i\tau u \, \mathbf{1}_{\{|u| \leq 1\}})\nu(du) \right\},$$

as required in (9.2). The increments of Y_ν have the form

$$Y_\nu(t_2) - Y_\nu(t_1) = \int_{\mathcal{R}} \mathbf{1}_{\{t_1 < s \leq t_2\}} \mathbf{1}_{\{|u| > 1\}} \, u \, N(du, ds)$$

$$+ \int_{\mathcal{R}} \mathbf{1}_{\{t_1 < s \leq t\}} \mathbf{1}_{\{|u| \leq 1\}} \, u \, \widetilde{N}(du, ds).$$

Since for non-overlapping intervals of time the supports of the corresponding integrands are disjoint (particularly, in variable s), the increments on non-overlapping intervals are independent.

Due to the shift-invariance of Lebesgue measure ds, the process Y_ν has stationary increments. Hence, it is a Lévy process. Finally by summing up independent Wiener and Poissonian parts, we see that

$$Y(t) := at + \sigma W(t) + Y_\nu(t)$$

is a required Lévy process satisfying (9.2).

Example 9.2 (Cauchy process). A Lévy process $Y = Y_\nu$ corresponding to $a = \sigma = 0$ and $\nu(du) = \frac{du}{\pi u^2}$ is called *Cauchy* process. We know from (2.25) that $Y(1)$ has the standard Cauchy distribution $\mathcal{C}(0, 1)$. It is easy to show (do it!) that $Y(t)$ has distribution $\mathcal{C}(0, t)$ for any $t > 0$.

Example 9.3 (Gamma process). Let $\Gamma(\cdot)$ denote the classical Gamma function,

$$\Gamma(\alpha) := \int_0^\infty x^{\alpha-1}e^{-x}dx, \qquad \alpha > 0.$$

The distribution $\mathcal{G}(\alpha)$ given by the density

$$p_\alpha(x) := \frac{x^{\alpha-1}e^{-x}}{\Gamma(\alpha)}, \qquad x > 0,$$

is called *Gamma distribution* with parameter α. In particular, $\alpha = 1$ corresponds to the well known *exponential distribution* with the density

$$p_1(x) = e^{-x}, \qquad x > 0,$$

and characteristic function

$$f_1(\tau) = \int_0^\infty e^{-(1-i\tau)x}dx = \frac{1}{1-i\tau}.$$

The family of Gamma distributions has a remarkable semi-group property. If X_1, X_2 are independent random variables with respective distributions $\mathcal{G}(\alpha_1)$, $\mathcal{G}(\alpha_2)$, then the distribution of $X_1 + X_2$ is $\mathcal{G}(\alpha_1 + \alpha_2)$. It follows immediately that all Gamma distributions are infinitely divisible and the corresponding characteristic functions are

$$f_\alpha(\tau) = \frac{1}{(1-i\tau)^\alpha}.$$

Let $\nu(du) := \frac{\mathbf{1}_{\{u>0\}}\,e^{-u}\,du}{u}$. Then, by using the well known formula for a definite integral, we have

$$\int_{\mathbb{R}} (e^{i\tau u} - 1)\nu(du) = \int_0^\infty \frac{e^{-(1-i\tau)u} - e^{-u}}{u}\,du = \ln\frac{1}{1-i\tau}.$$

We see that ν is the intensity measure corresponding to the exponential distribution. Remark that

$$\int_{\mathbb{R}} u\mathbf{1}_{\{|u|\leq 1\}}\nu(du) = \int_0^1 e^{-u}du = 1 - e^{-1}.$$

Therefore, we may write in our standard notation,

$$\ln f_1(\tau) = \ln\frac{1}{1-i\tau} = i(1-e^{-1})\tau + \int_{\mathbb{R}} (e^{i\tau u} - 1 - i\tau u\mathbf{1}_{\{|u|\leq 1\}})\,\nu(du).$$

Thus the triplet of exponential distribution writes as $\sigma^2 = 0$, $a = 1 - e^{-1}$, $\nu(du) = \frac{\mathbf{1}_{\{u>0\}}\,e^{-u}\,du}{u}$. The Lévy process Y corresponding to this triplet is called *Gamma process*.

Exercise 9.4. Prove that for each $t > 0$ the distribution of Gamma process $Y(t)$ is $\mathcal{G}(t)$.

9.2 Compound Poisson processes

Recall that a compound Poisson random variable associated with a finite measure ν on $\mathbb{R}\backslash\{0\}$ was introduced in (2.5) as $S := \sum_{j=1}^{N} U_j$, where (U_j) is an i.i.d. sequence of random variables with common distribution $\frac{\nu}{|\nu|}$, and N is a Poisson random variable with distribution $P_N = \mathcal{P}(|\nu|)$, independent of (U_j).

We may further develop this construction by introducing dynamics in time. Let now $N(t), t \geq 0$, be a Poisson process of intensity $|\nu|$ independent of (U_j) and define *compound Poisson process* of intensity ν by

$$X(t) := \sum_{j=1}^{N(t)} U_j. \tag{9.4}$$

If we consider X as a (random) function of time, we see a step function with a jump U_j at the instant of j-th jump of Poisson process N.

By proceeding as in (2.6) it is easy to see that for any $0 \leq t_1 \leq t_2$ and any $\tau \in \mathbb{R}$

$$\mathbb{E}\, e^{i\tau(X(t_2)-X(t_1))} = \exp\left\{(t_2 - t_1) \int (e^{i\tau u} - 1)\nu(du)\right\}.$$

This turns out to be a special case of (9.2) if we let there $\sigma^2 = 0$ and $a = \int u \mathbf{1}_{\{|u|\leq 1\}}\nu(du)$. Hence, X is a process with stationary increments distributed exactly as it should be for a Lévy process of intensity ν. It is not difficult to show that the increments of X are independent. Thus compound Poisson processes form a subclass of Lévy processes.

Notice that neither Gamma process nor Cauchy process belong to this subclass because the corresponding intensity measures are infinite.

9.3 Stable Lévy processes

Since stable distributions are infinitely divisible, we may consider the corresponding subclass of Lévy processes. Thus a Lévy process Y is called α-stable (resp. *strictly α-stable*), if $Y(1)$ is an α-stable (resp. *α-strictly stable*) random variable. We know that in this case $Y(t)$ is a stable (strictly stable) variable for any $t \geq 0$.

A typical trajectory of a stable Lévy process is represented on Figure 9.1. Vertical segments correspond to the jumps of the process.

For example, Wiener process is a strictly 2-stable Lévy process, Cauchy process is a strictly 1-stable Lévy but Poisson and Gamma processes do not belong to that class.

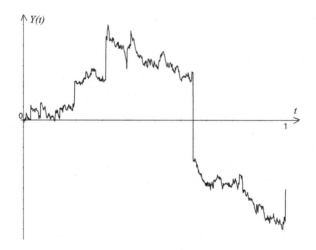

Fig. 9.1   Trajectory of a stable Lévy process.

Exercise 9.5. Prove that any α-stable Lévy process is an α-stable process according to the definition given in Section 5. This means that for any $0 < t_1 < \cdots < t_n$ the random vector $Z := (Y(t_1), \ldots, Y(t_n))$ is α-stable, or, equivalently, for any $v \in \mathbb{R}^n$ the scalar product $(v, Z) = \sum_{j=1}^n v_j Y(t_j)$ is an α-stable random variable.

Strictly stable Lévy processes are particularly interesting because, in addition to independence and stationarity of increments, they are self-similar.

For notation simplicity, let us exclude the case $\alpha = 1$ for a while, although the results exposed below remain valid in that case, too. So let $\alpha \neq 1$. Recall that $\widetilde{S}(c_-, c_+, \alpha)$ denotes the distribution of a random variable with log-characteristic function (7.28). Let now Y be an strictly α-stable Lévy process. Then $P_{Y(1)} = \widetilde{S}(c_-, c_+, \alpha)$ for appropriate parameters c_-, c_+. It follows from the triplet considerations for general Lévy processes that for any $t \geq 0$ we have

$$P_{Y(t)} = \widetilde{S}(c_- t, c_+ t, \alpha). \tag{9.5}$$

We are now able to show that Y is H-self-similar with $H = \frac{1}{\alpha}$. Take any $c > 0$ and consider the scaled process $Y_c(t) := c^{-1/\alpha} Y(ct)$. It is obvious that Y_c is a Lévy process. Moreover, $Y_c(1) := c^{-1/\alpha} Y(c)$. By using (9.5) and the scaling rule (7.29) we conclude that $Y_c(1)$ has the same distribution $\widetilde{S}(c_-, c_+, \alpha)$ as $Y(1)$. As we know, the distribution of a Lévy process is uniquely determined by the distribution of its value at time one. Hence the processes $Y_c(\cdot)$ and $Y(\cdot)$ have the same finite-dimensional distributions. This is exactly the property required for H-self-similarity.

In case $\alpha = 1$ we obtain 1-self-similarity by the same arguments after replacing generic distribution $\widetilde{\mathcal{S}}(c_-, c_+, \alpha)$ with its analogue $\widetilde{\mathcal{S}}(a, c, 1)$.

An important class of strictly stable processes may be constructed by considering *exit times*. Let W be a Wiener process. Let the exit time

$$Y(r) := \inf\{t : W(t) = r\}, \qquad r \geq 0,$$

be the first time instant when the trajectory of W reaches the level r. The random process Y is increasing and it inherits self-similarity from W. More precisely, let $c > 0$. Consider a Wiener process $W_c(\cdot) := \frac{W(c^2 \cdot)}{c}$ and the corresponding exit times $Y_c(\cdot)$. It is obvious that

$$Y_c(\cdot) = c^{-2} Y(c \cdot).$$

Since $Y_c(\cdot)$ is equidistributed with $Y(\cdot)$, it follows that $Y(\cdot)$ is 2-self-similar (recall that W is $\frac{1}{2}$-self-similar).

It is much less obvious that Y is a Lévy process. Recall that for any $t_0 \geq 0$ the process $W(t_0 + t) - W(t)$ is again a Wiener process independent on the past $\{W(s), s \leq t\}$, i.e. at any instant t_0 Wiener process restarts its motion from scratch. One can show (we do not prove this deeper fact here!) that the same happens at random instant when W reaches some level. In other words, the process

$$\widetilde{W}_r(t) := W(Y(r) + t) - W(Y(r)) = W(Y(r) + t) - r$$

is again a Wiener process. Let $\widetilde{Y}_r$ be the corresponding exit times,

$$\widetilde{Y}_r(r') := \inf\{t : \widetilde{W}_r(t) = r'\}, \qquad r' \geq 0.$$

Given $r_1, r_2 \geq 0$, it takes some time for Wiener process to reach the level r_1 and then some more time to climb from level r_1 to the higher level $r_1 + r_2$. Thus we have identity

$$Y(r_1 + r_2) = Y(r_1) + \widetilde{Y}_{r_1}(r_2). \tag{9.6}$$

Reading this as $Y(r_1 + r_2) - Y(r_1) = \widetilde{Y}_{r_1}(r_2)$ and recalling that $\widetilde{Y}_{r_1}(\cdot)$ is equidistributed with $Y(\cdot)$ we see that Y is a process with stationary increments. Moreover, since $\widetilde{W}_{r_1}(\cdot)$ (hence also $\widetilde{Y}_{r_1}(\cdot)$) does not depend on the behavior of W prior to the exit time $Y(r_1)$, we infer that the increment $Y(r_1 + r_2) - Y(r_1)$ is independent of $\{Y(r), r \leq r_1\}$ and conclude that Y is a process with independent increments.

Taking together the self-similarity and independence of increments we may rewrite (9.6) as

$$(r_1 + r_2)^2 Y = r_1^2 Y' + r_2^2 Y''$$

where Y, Y', Y'' are equidistributed with $Y(1)$ and Y', Y'' are independent. This simply means that $Y(1)$ is a $\frac{1}{2}$-stable variable.

Exceptionally, the distribution of $Y(1)$ may be written explicitly due to the known formula for the distribution of supremum of Wiener process. It is known that

$$\mathbb{P}\left(\sup_{0 \le t \le 1} W(t) \ge r \right) = 2\,\mathbb{P}\left(W(1) \ge r \right) = 2\left(1 - \Phi(r) \right).$$

Self-similarity of W transforms this fact into

$$\mathbb{P}\left(\sup_{0 \le t \le T} W(t) \ge r \right) = 2\,\mathbb{P}\left(W(T) \ge r \right) = 2\left(1 - \Phi(r/\sqrt{T}) \right),$$

for any $T > 0$. Hence,

$$\mathbb{P}\left(Y(1) > T \right) = \mathbb{P}\left(\sup_{0 \le t \le T} W(t) < 1 \right) = 2\Phi(1/\sqrt{T}) - 1.$$

Differentiation yield a formula for the distribution density of $Y(1)$,

$$p_{Y(1)}(T) = \frac{2\Phi'(1/\sqrt{T})}{2T^{3/2}} = \frac{\exp\{-\frac{1}{2T}\}}{\sqrt{2\pi}T^{3/2}}.$$

When T goes to zero, the density is decreasing exponentially fast. This means that Wiener process is almost unable to reach the unit level in short time. On the contrary, when T goes to infinity, the density is decreasing rather slowly: if W makes a long excursion into the negative half-line, reaching the unit level may take a long time.

The exit time construction presented here may be extended by replacing Wiener process with other strictly α-stable Lévy processes *without positive jumps*. Indeed, if one allows positive jumps, then a level r_1 is attained with an overshot jump and the left hand side of (9.6) turns out to be smaller than the right hand side. The corresponding exit time process Y will be increasing α-self similar, and $\frac{1}{\alpha}$-stable Lévy process.

In this section we merely discussed the notion of Lévy process. The true deep theory can be found in monographs of J. Bertoin [7], K. Sato [90] and in the lectures of A. Kyprianou [55].

10 Spectral Representations for Stationary Processes

10.1 *Wide sense stationary processes*

Recall that $Y(t), t \in \mathbb{R}$, is called a stationary process if for any real numbers $s, t_1, \ldots, t_n$ we have

$$P^Y_{t_1,\ldots,t_n} = P^Y_{s+t_1,\ldots,s+t_n}.$$

In particular, for $n = 1, t_1 = 0$ we have $P_{Y(s)} = P_{Y(0)}$ for any $s \in \mathbb{R}$. By letting $n = 2$ we see that the two-dimensional vectors $(Y(t_1), Y(t_2))$ and $(Y(t_1 - t_2), Y(0))$ are equidistributed. These equalities of distributions lead to the equalities for expectations and covariances. Namely, if $a := \mathbb{E}\,Y(0)$ is finite, then we have

$$\mathbb{E}\,Y(t) = a, \qquad t \in \mathbb{R}.$$

If $\sigma^2 := \mathbb{E}\,Y(0)^2$ is finite, then

$$\mathbb{E}\,Y(t)^2 = \sigma^2, \qquad t \in \mathbb{R}.$$

Moreover, in this case the covariance

$$K(t) := \text{cov}(Y(t), Y(0)), \qquad t \in \mathbb{R},$$

is well defined and we have

$$K_Y(t_1, t_2) = \text{cov}(Y(t_1), Y(t_2)) = \text{cov}(Y(t_1 - t_2), Y(0)) = K(t_1 - t_2).$$

These relations are already sufficient for proving many important properties of the process Y. Therefore, we may introduce the following definition.

A random process $Y(t), t \in \mathbb{R}$, is called a *wide sense stationary process* if $\mathbb{E}\,Y(t)^2 < \infty$ for all $t \in T$ and

$$\mathbb{E}\,Y(t) = a, \qquad t \in \mathbb{R},$$

$$K_Y(t_1, t_2) = \text{cov}(Y(t_1), Y(t_2)) = K(t_1 - t_2), \qquad t_1, t_2 \in \mathbb{R},$$

for some constant a and for some function $K(\cdot)$.

It turns out to be convenient to consider first *complex-valued* processes, in which case $a \in \mathbb{C}$ and $K(\cdot)$ is a complex-valued function.

As we have seen, every stationary process with finite variance is a wide sense stationary process but the converse is of course not true. Notice however the following important special case.

Exercise 10.1. If Y is a Gaussian wide sense stationary process, then it is also a stationary process.

The covariance function $K(\cdot)$ is non-negative definite in the sense of (1.15), since for all $n \in \mathbb{N}$, $t_1, \ldots, t_n \in \mathbb{R}$, and $c_1, \ldots, c_n \in \mathbb{C}$ we have

$$\sum_{j,k=1}^{n} K(t_j - t_k) c_j \overline{c_k} = \sum_{j,k=1}^{n} \mathrm{cov}(Y(t_j), Y(t_k)) c_j \overline{c_k}$$

$$= \mathrm{cov}\left(\sum_{j=1}^{n} c_j Y(t_j), \sum_{k=1}^{n} c_k Y(t_k) \right)$$

$$= \mathbb{E} \left| \sum_{j=1}^{n} c_j (Y(t_j) - \mathbb{E}Y(t_j)) \right|^2 \geq 0.$$

From now on, we assume a mild continuity condition on $K(\cdot)$. Namely, let $K(\cdot)$ be continuous at zero. Then our process Y is $\mathbb{L}_2$-continuous, since for any $t \in \mathbb{R}$

$$\mathbb{E} |Y(t+s) - Y(t)|^2 = \mathbb{E} |Y(s) - Y(0)|^2$$
$$= \mathrm{cov}(Y(s) - Y(0), Y(s) - Y(0))$$
$$= \mathrm{cov}(Y(s), Y(s)) + \mathrm{cov}(Y(0), Y(0)) - \mathrm{cov}(Y(s), Y(0))$$
$$-\mathrm{cov}(Y(0), Y(s))$$
$$= 2K(0) - K(s) - \overline{K(s)} = 2(K(0) - ReK(s))$$

goes to zero as s goes to zero.

Exercise 10.2. Prove that Y is $\mathbb{L}_2$-continuous if and only if $K(\cdot)$ is continuous at zero.

10.2 *Spectral representations*

If $K(\cdot)$ is both non-negative and continuous, then by Bochner–Khinchin Theorem stated in Subsection 1.8, there exists a unique finite measure F on $\mathbb{R}$ such that

$$K(\tau) = \int_{-\infty}^{\infty} e^{i\tau u} F(du), \qquad \tau \in \mathbb{R}. \tag{10.1}$$

The measure F is called the *spectral measure* for the process Y. In particular, we have

$$\mathbb{V}arY(t) = K(0) = F(\mathbb{R}) \qquad \text{for all } t \in \mathbb{R}.$$

If the measure F has a density $f(\cdot)$, i.e. $F(du) = f(u)du$, then f is called a *spectral density* for Y.

We will show now how integration over complex measures with uncorrelated values enables to construct wide sense stationary processes. Namely, let X be a centered complex-valued measure with uncorrelated values on $\mathbb{R}$ with intensity measure F. Take any $a \in \mathbb{C}$ and let

$$Y(t) := a + \int_{-\infty}^{\infty} e^{itu} X(du), \qquad t \in \mathbb{R}. \tag{10.2}$$

Then Y is a wide sense stationary processes because clearly $\mathbb{E}\, Y(t) = a$ and by (7.10) we have

$$\mathrm{cov}(Y(t_1), Y(t_2)) = \mathbb{E} \left(\int_{\mathbb{R}} e^{it_1 u} X(du), \overline{\int_{\mathbb{R}} e^{it_2 u} X(du)} \right)$$

$$= \int_{\mathbb{R}} e^{i(t_1 - t_2)u} F(du) = K(t_1 - t_2),$$

where $K(\cdot)$ is connected to F via formula (10.1). Clearly, F turns out to be a spectral measure for Y. Formulae (10.1) and (10.2) are called *spectral representations* for K and Y, respectively.

One may prove that *every* square mean continuous process that is wide sense stationary admits a spectral representation (10.2) with appropriate F and X.

To give a feeling of work with spectral representations let us consider differentiation of a stationary process Y given by (10.2). Formally,

$$Y'(t) := \lim_{s \to 0} \frac{Y(t+s) - Y(t)}{s}$$

$$= \lim_{s \to 0} \int_{\mathbb{R}} \frac{e^{i(t+s)u} - e^{itu}}{s} \, X(du)$$

$$= \lim_{s \to 0} \int_{\mathbb{R}} e^{itu} \frac{e^{isu} - 1}{s} \, X(du)$$

$$= \int_{\mathbb{R}} e^{itu} \lim_{s \to 0} \frac{e^{isu} - 1}{s} \, X(du)$$

$$= \int_{\mathbb{R}} e^{itu} \, iu \, X(du).$$

Recall that the latter integral is well defined iff

$$\int_{\mathbb{R}} u^2 F(du) < \infty.$$

This is indeed a necessary and sufficient condition for $\mathbb{L}_2$-differentiability, since if it is satisfied, the boundedness of the fraction $\frac{e^{isu} - 1}{su}$ and Lebesgue dominated convergence theorem yield,

$$\lim_{s \to 0} \mathbb{E} \left(\frac{Y(t+s) - Y(t)}{s} - Y'(t) \right)^2$$

$$= \lim_{s \to 0} \int_{\mathbb{R}} \left| e^{itu} \left(\frac{e^{isu} - 1}{isu} - 1 \right) \right|^2 u^2 F(du) = 0.$$

By iterating this procedure, we find that the m-th square mean derivative of Y exists iff

$$\int_{\mathbb{R}} u^{2m} F(du) < \infty.$$

If this condition holds, then

$$Y^{(m)}(t) = \int_{\mathbb{R}} e^{itu} (iu)^m X(du).$$

We see that the derivative $Y^{(m)}$ also is a wide sense stationary process.

Exercise 10.3. Describe a random measure with uncorrelated values X_m such that the spectral representation for $Y^{(m)}$ holds, i.e.

$$Y^{(m)}(t) = \int_{\mathbb{R}} e^{itu} X_m(du).$$

Prove that the measure $F_m(du) := u^{2m} F(du)$ is the intensity measure for X_m and the spectral measure for $Y^{(m)}$.

Let us consider the *law of large numbers* (LLN) as another application of spectral representation.

Proposition 10.4 (LLN for stationary processes). *Let Y be a wide sense stationary process admitting the spectral representation* (10.2). *Then*

$$\frac{1}{T} \int_0^T Y(t)dt \overset{L_2}{\to} a + X(\{0\}), \qquad as \ T \to \infty.$$

In particular, if the corresponding spectral measure F satisfies $F(\{0\}) = 0$, then it is true that

$$\frac{1}{T} \int_0^T Y(t)dt \overset{L_2}{\to} a, \qquad as \ T \to \infty.$$

Proof. Without loss of generality we may assume $a = 0$. Let

$$A_T := \frac{1}{T} \int_0^T Y(t)dt.$$

Then

$$A_T = \frac{1}{T} \int_0^T \left(\int_{\mathbb{R}} e^{itu} X(du) \right) dt$$

$$= \int_{\mathbb{R}} \left(\frac{1}{T} \int_0^T e^{itu} dt \right) X(du)$$

$$= \int_{\mathbb{R}} \left(\mathbf{1}_{\{0\}}(u) + \mathbf{1}_{\{\mathbb{R}\setminus 0\}}(u) \frac{e^{iTu} - 1}{iTu} \right) X(du).$$

Notice that

$$X(\{0\}) = \int_{\mathbb{R}} \mathbf{1}_{\{0\}}(u)\, X(du).$$

By using isometric property of the integral (7.9) and Lebesgue dominated convergence theorem, we have

$$\mathbb{E}\, (A_T - X(\{0\}))^2 = \int_{\mathbb{R}\backslash 0} \left| \frac{e^{iTu} - 1}{iTu} \right|^2 F(du) \to 0,$$

as T goes to infinity. The first claim of proposition is proved.

Since $\mathbb{E}\, X(\{0\}) = 0$, if $\mathbb{V}ar X(\{0\}) = F(\{0\}) = 0$, we have $X(\{0\}) = 0$, and the second claim of proposition follows from the first one. $\square$

Let us now consider the case of *real-valued* stationary processes. It seems natural to search their representations via *real-valued* measures with uncorrelated values.

Let F be a finite measure on $[0, \infty)$. Let $X^{(re)}, X^{(im)}$ be two real-valued mutually uncorrelated random measures with uncorrelated values on $(0, \infty)$, each having intensity $F/2$. Let X_0 be a centered real-valued random variable with variance $F(\{0\})$. We assume that X_0 is uncorrelated with $X^{(re)}, X^{(im)}$.

We define now the total complex-valued random measure X on $\mathbb{R}$, first by

$$X(A) := \begin{cases} X^{(re)}(A) + iX^{(im)}(A), & A \subset (0, \infty), \\ X^{(re)}(-A) - iX^{(im)}(-A), & A \subset (-\infty, 0), \end{cases}$$

and, in general case $A \in \mathcal{B}$,

$$X(A) := X(A \cap (-\infty, 0)) + X_0 \mathbf{1}_{\{0 \in A\}} + X(A \cap (0, \infty))$$

It is easy to check that X is a complex-valued random measure with dependent but uncorrelated values. The dependence shows up in the relation $X(-A) = \overline{X(A)}$ for any measurable A, while the absence of correlation is based upon the following computation. Let $A \subset (-\infty, 0)$, $B \subset (0, \infty)$. Then

$$\begin{aligned}
&\mathrm{cov}(X(A), X(B)) \\
&= \mathrm{cov}(X^{(re)}(-A) - iX^{(im)}(-A), X^{(re)}(B) + iX^{(im)}(B)) \\
&= \mathrm{cov}(X^{(re)}(-A), X^{(re)}(B)) + (-i)\,\overline{i}\,\mathrm{cov}(X^{(im)}(-A), X^{(im)}(B)) \\
&= F((-A) \cap B)/2 + (-i)^2 F((-A) \cap B)/2 = 0.
\end{aligned}$$

Notice that the intensity measure of X is equal to symmetric extension of F, where we let $F(-A) := F(A)$ for any measurable $A \subset (0, \infty)$; furthermore, we have

$$\mathbb{E}\,|X(A)|^2 = \mathbb{E}\,X^{(re)}(A)^2 + \mathbb{E}\,X^{(im)}(A)^2 = F(A)/2 + F(A)/2 = F(A).$$

Let us now take any $a \in \mathbb{R}$ and consider a stationary process defined by the spectral representation (10.2). Then Y is real-valued and can be represented through initial real-valued random measures as follows

$$Y(t) = a + \int_0^\infty e^{itu}(X^{(re)} + iX^{(im)})(du) + X_0$$

$$+ \int_0^\infty e^{-itu}(X^{(re)} - iX^{(im)})(du)$$

$$= a + X_0 + 2\int_0^\infty \cos(tu)\,X^{(re)}(du) \qquad (10.3)$$

$$-2\int_0^\infty \sin(tu)\,X^{(im)}(du).$$

Notice that all parts of representation are mutually uncorrelated.

This representation works for *arbitrary* real-valued wide sense stationary process. However, it is often more convenient to express the real-valued process Y via complex-valued measure X, as done in (10.2).

Example 10.5 (Two-point spectrum). Let the random measures $X^{(re)}$ and $X^{(im)}$ from the previous construction be concentrated at a single common point $h > 0$, i.e.

$$X^{(re)}(A) := X_1 \mathbf{1}_{\{h \in A\}}, \qquad A \subset (0, \infty);$$
$$X^{(im)}(A) := X_2 \mathbf{1}_{\{h \in A\}}, \qquad A \subset (0, \infty),$$

where X_1, X_2 are uncorrelated centered real-valued random variables with finite variance. Assume additionally that

$$\mathbb{V}ar X_1 = \mathbb{V}ar X_2 := \sigma^2/2.$$

Then, taking symmetrization into account, the spectral measure F of Y has weights σ^2 at points $\pm h$, whereas the name "two-point spectrum". Now

$$Y(t) = a + 2\cos(th)X_1 - 2\sin(th)X_2 = a + A\cos(th + \phi),$$

turns out to be a periodic motion with random amplitude $A = 2\sqrt{X_1^2 + X_2^2}$, deterministic frequency h, and random phase ϕ being a solution of equations

$$\cos\phi = \frac{X_1}{\sqrt{X_1^2 + X_2^2}},$$
$$\sin\phi = \frac{X_2}{\sqrt{X_1^2 + X_2^2}}.$$

Exercise 10.6. Let X be a real-valued centered random measure on $\mathbb{R}$ with uncorrelated values and intensity equal to Lebesgue measure. Let $h \in \mathbb{L}_2(\mathbb{R})$. Consider a random process

$$Y(t) := \int_{\mathbb{R}} h(u - t) X(du). \qquad (10.4)$$

One may call this formula a *shift representation* for Y. Prove that Y is a wide sense stationary process. Find its spectral density in terms of the Fourier transform of h.

Example 10.7 (Ornstein–Uhlenbeck process). In Subsection 6.3 we introduced Ornstein–Uhlenbeck process having covariance function

$$K(t) = e^{-|t|/2}.$$

Looking at the formula (1.18) for characteristic function of Cauchy distribution $\mathcal{C}(0, 1/2)$

$$e^{-|t|/2} = \int_{\mathbb{R}} \frac{2e^{itu} du}{\pi(1 + 4u^2)}$$

from the different point of view, we conclude that the spectral density of Ornstein–Uhlenbeck process is

$$f(u) = \frac{2}{\pi(1 + 4u^2)} \ .$$

Let now X be a Gaussian white noise with Lebesgue intensity. Choose

$$h(u) = e^{-u/2} \, \mathbf{1}_{\{u>0\}} \ .$$

In agreement with (10.4), consider the corresponding shift representation

$$U(t) := \int_t^{\infty} e^{-(u-t)/2} \, X(du).$$

It is clear that U is a centered Gaussian process. Moreover, by using isometric property of integrals, for $t \geq 0$ we obtain

$$
\begin{aligned}
K_U(t) &= \mathbb{E}\, U(t) U(0) \\
&= \mathbb{E} \left(\int_t^{\infty} e^{-(u-t)/2} \, X(du) \cdot \int_0^{\infty} e^{-u/2} \, X(du) \right) \\
&= \int_t^{\infty} e^{-(u-t)/2} \cdot e^{-u/2} \, du \\
&= e^{-t/2} \int_t^{\infty} e^{-(u-t)} \, du = e^{-t/2}.
\end{aligned}
$$

It follows that U is an Ornstein–Uhlenbeck process.

10.3   *Further extensions*

The spectral representations briefly described in this section admit various extensions, e.g. for the cases of random fields $(t, u \in \mathbb{R}^d)$, for random sequences $(t \in \mathbb{Z}, u \in \mathbb{S}^1)$, or for periodic processes $(t \in \mathbb{S}^1, u \in \mathbb{Z})$, where $\mathbb{Z}$ is the set of integer numbers and $\mathbb{S}^1$ denotes the unit circle.

A random sequence $Y(n), n \in \mathbb{Z}$, complex- or real-valued, is called a *wide sense stationary sequence* if $\mathbb{E}\, Y(n)^2 < \infty$ for all $n \in \mathbb{Z}$ and

$$\mathbb{E}\, Y(n) = a, \qquad n \in \mathbb{Z},$$

$$K_Y(n_1, n_2) = \operatorname{cov}(Y(n_1), Y(n_2)) = K(n_1 - n_2), \qquad n_1, n_2 \in \mathbb{Z},$$

for some constant a and for some function $K(\cdot)$.

Any complex-valued wide sense stationary sequence admits a representation

$$Y(n) = a + \int_{\mathbb{S}^1} u^n X(du), \qquad n \in \mathbb{Z}, \tag{10.5}$$

where X is a centered complex-valued measure with uncorrelated values on the unit circle $\mathbb{S}^1$.

Exercise 10.8. Let $Y(t), t \in \mathbb{R}$ be a wide sense stationary process having the spectral representation (10.2). Prove that its restriction on $\mathbb{Z}$ is a wide sense stationary sequence and find its spectral representation (10.5).

A random field $Y(t), t \in \mathbb{R}^d$, complex- or real-valued, is called a *wide sense stationary random field* if $\mathbb{E}\, Y(t)^2 < \infty$ for all $t \in \mathbb{R}^d$ and

$$\mathbb{E}\, Y(t) = a, \qquad t \in \mathbb{R}^d,$$

$$K_Y(t_1, t_2) = \operatorname{cov}(Y(t_1), Y(t_2)) = K(t_1 - t_2), \qquad t_1, t_2 \in \mathbb{R}^d,$$

for some constant a and for some function $K(\cdot)$.

Any complex-valued wide sense stationary random field admits a representation

$$Y(t) = a + \int_{\mathbb{R}^d} e^{i(t,u)} X(du), \qquad t \in \mathbb{R}^d, \tag{10.6}$$

where X is a centered complex-valued measure with uncorrelated values on $\mathbb{R}^d$.

Exercise 10.9. Prove that any random field $Y(t), t \in \mathbb{R}^d$, admitting a representation (10.6) is a wide sense stationary random field.

A random process $Y(t), t \in \mathbb{R}$, complex- or real-valued, is called *periodic* with period T if

$$Y(t+T) = Y(t), \qquad t \in \mathbb{R}.$$

If a process Y is wide sense stationary *and* periodic with period T, its spectral measure F is concentrated on the grid $\{hn, n \in \mathbb{Z}\}$ where $h := \frac{2\pi}{T}$ is called the *frequency* of the process. In other words, Y admits a series representation

$$Y(t) = a + \sum_{n \in \mathbb{Z}} X_n \, e^{2\pi i t/T}, \qquad t \in \mathbb{R},$$

where $(X_n)_{n \in \mathbb{Z}}$, is a sequence of centered uncorrelated complex-valued random variables such that $\sum_{n \in \mathbb{Z}} \mathbb{E} X_n^2 < \infty$.

Moreover, if Y is real-valued, then its real representation (10.3) also becomes a series representation

$$Y(t) = a + X_0 + \sum_{n=1}^{\infty} \left(X_n' \cos(2\pi t/T) + X_n'' \sin(2\pi t/T) \right), \qquad t \in \mathbb{R},$$

where $X_0, (X_n')_{n \in \mathbb{N}}, (X_n'')_{n \in \mathbb{N}}$, are centered uncorrelated real-valued random variables and $\mathbb{E} X_n'^2 = \mathbb{E} X_n''^2$ for each $n \in \mathbb{N}$.

The spectral representation can be also extended to random processes (or fields) with wide sense stationary increments.

A process $Y(t), t \in T$, where $T = \mathbb{R}$ or $T = \mathbb{R}_+$, is called a *process with wide sense stationary increments* if for all $s \in T$ the processes Y_s defined by $Y_s(t) := Y(s+t) - Y(s)$ have the same expectations and covariance functions. For such processes, the analogue of spectral representation (10.2) is

$$Y(t) := a + X_0 + X_1 t + \int_{\mathbb{R}} (e^{itu} - 1) X(du), \qquad t \in \mathbb{R}, \qquad (10.7)$$

and the intensity measure F corresponding to X, which is eventually infinite, must satisfy Lévy–Khinchin condition

$$\int_{\mathcal{R}} \min\{u^2, 1\} F(du) < \infty.$$

In particular, if Y is a wide sense stationary process, then the increments $\widetilde{Y}(t) := Y(t) - Y(0)$ form a process with wide sense stationary increments and random measure X in (10.2) for Y and in (10.7) for $\widetilde{Y}$ is the same while X_0 and X_1 in (10.7) vanish.

See [106, 113] for a more detailed exposition of the theory of spectral representations and its applications.

11   Convergence of Random Processes

11.1   *Finite-dimensional convergence*

We say that a sequence of processes $X_m(t), t \in T$, *converges to a process*
$X(t), t \in T$, *in finite-dimensional distributions* (f.d.d.) if for any $n \geq 1$ and
any $t_1, \ldots, t_n$ in T we have the weak convergence of distributions

$$P^{X_m}_{t_1,\ldots,t_n} \Rightarrow P^{X}_{t_1,\ldots,t_n}.$$

In this case we write $X_m \xrightarrow{\text{f.d.d.}} X$.

If $g : \mathbb{R}^n \to \mathbb{R}^1$ is a continuous function, then by Exercise 4.3 we have

$$g(X_m(t_1), \ldots, X_m(t_n)) \Rightarrow g(X(t_1), \ldots, X(t_n)). \tag{11.1}$$

For example,

$$\max_{1 \leq j \leq n} X_m(t_j) \Rightarrow \max_{1 \leq j \leq n} X(t_j),$$

$$\sum_{j=1}^{n} X_m(t_j)^2 \Rightarrow \sum_{j=1}^{n} X(t_j)^2,$$

etc.

Let us now consider three typical examples of convergence.

Proposition 11.1. *Let $X_m(t), t \geq 0$, be Poisson processes of intensity m
and let $Z_m(t) := m^{-1/2}(X_m(t) - mt)$ denote their centered and rescaled
versions. Then, as $m \to \infty$,*

$$Z_m \xrightarrow{\text{f.d.d.}} W,$$

where W is a Wiener process.

Proof.   We have to check that for any $n \geq 1$ and any $0 \leq t_1 < \cdots < t_n$
it is true that $P^{Z_m}_{t_1,\ldots,t_n} \Rightarrow P^{W}_{t_1,\ldots,t_n}$. We know from Exercise 4.7 that this is
equivalent (taking into account that $Z_m(0) = W(0) = 0$) to convergence of
differences

$$(Z_m(t_1) - Z_m(0), Z_m(t_2) - Z_m(t_1), \ldots, Z_m(t_n) - Z_m(t_{n-1}))$$
$$\Rightarrow (W(t_1) - W(0), W(t_2) - W(t_1), \ldots, W(t_n) - W(t_{n-1})).$$

Next, recall that both Poisson and Wiener processes have independent increments. Therefore, by Exercise 4.6 it is sufficient to check one-dimensional
convergence

$$Z_m(t) - Z_m(s) \Rightarrow W(t) - W(s)$$

for any $t > s \geq 0$. Notice that by definition of Poisson process and by definition of Z_m, the variable $Z_m(t) - Z_m(s)$ is equidistributed with $m^{-1/2}\,(Y_m - m(t - s))$ where Y_m is a Poisson random variable of intensity $m(t - s)$. Furthermore, by Exercise 1.3, Y_m is equidistributed with $\sum_{j=1}^{m} Y^{(j)}$ where $(Y^{(j)})_{1 \leq j \leq m}$ are independent $\mathcal{P}(t - s)$-distributed random variables. It follows that Z_m is equidistributed with $Z'_m :=$ $m^{-1/2}\sum_{j=1}^{m}(Y^{(j)} - (t - s))$. By Lévy's Central Limit Theorem (cf. Section 3) the distribution of $\frac{Z'_m}{\sqrt{t-s}}$ converges to $\mathcal{N}(0, 1)$. Hence the distribution of Z_m converges to $\mathcal{N}(0, t - s) = P_{W(t)-W(s)}$. $\qquad\square$

Exercise 11.2. Prove the following extension of Proposition 11.1. Let $X(t), t \geq 0$, be a Lévy process such that $\mathbb{E}\,X(1) = 0$ and $\mathbb{E}\,X(1)^2 = 1$. Let $Z_m(t) := m^{-1/2}X(mt)$. Then, as $m \to \infty$,

$$Z_m \xrightarrow{\text{f.d.d.}} W,$$

where W is a Wiener process.

The next example represents a simplest form of so called *invariance principle*. Let $X_1, X_2, \ldots$ be a sequence of independent identically distributed random variables. Then the sequence of sums $S_0 := 0$ and

$$S_n := \sum_{j=1}^{n} X_j, \qquad n \geq 1,$$

is called a *random walk* with steps X_j. We formally render it continuous-time process by letting

$$S(t) := S_n, \qquad n \leq t < n + 1, \ n \geq 0.$$

The word "invariance" means that in the long run the behaviour of $S(\cdot)$ does not depend on the distribution of steps X_j under quite mild assumptions on this distribution.

Proposition 11.3 (*f.d.d.*-invariance principle). *Assume that the steps of a random walk satisfy* $\mathbb{E}\,X_j = 0$ *and* $0 < \mathbb{E}\,X_j^2 := \sigma^2 < \infty$. *Let*

$$Z_m(t) := \frac{S(mt)}{\sigma\sqrt{m}}, \qquad t \geq 0.$$

Then, as $m \to \infty$,

$$Z_m \xrightarrow{\text{f.d.d.}} W,$$

where W is a Wiener process.

Proof. Again we handle the processes with independent increments. The arguments of the the previous proof reduce our goal to checking one-dimensional convergence

$$Z_m(t) - Z_m(s) \Rightarrow W(t) - W(s)$$

for any $t > s \geq 0$. From the definitions of $S(\cdot)$ and $Z_m(\cdot)$ it follows that

$$Z_m(t) - Z_m(s) = \frac{S([mt]) - S([ms])}{\sigma\sqrt{m}}$$

where $[x]$ denotes the integer part of $x \geq 0$, i.e. the largest integer not exceeding x. Rewrite the fraction as a product

$$Z_m(t) - Z_m(s) = \frac{\sum_{j=[ms]+1}^{[mt]} X_j}{\sigma\sqrt{[mt] - [ms]}} \; \frac{\sqrt{[mt] - [ms]}}{\sqrt{m}}.$$

By Lévy's Central Limit Theorem the distribution of the first factor converges to the standard normal law $\mathcal{N}(0,1)$, while the second term is deterministic and tends to $\sqrt{t - s}$. Hence, the distribution of $Z_m(t) - Z_m(s)$ converges to $\mathcal{N}(0, t - s) = P_{W(t)-W(s)}$. $\qquad\square$

Exercise 11.4. State and prove an analogue of Proposition 11.3 for a random walk with steps that belong to the domain of attraction of an α-stable distribution with $\alpha < 2$. Hint: the limiting process is an α-stable Lévy process.

Remark 11.5. Convergence $Z_m \overset{\text{f.d.d.}}{\longrightarrow} W$ implies $\frac{S_m}{\sigma\sqrt{m}} = Z_m(1) \Rightarrow W(1)$. Since $P_{W(1)} = \mathcal{N}(0,1)$, Proposition 11.3 includes Lévy's Central Limit Theorem.

The third basic example comes from Statistics. Let $X_1, X_2, \ldots$ be i.i.d. random variables uniformly distributed on the interval $[0,1]$, i.e. their common distribution function is

$$F_{X_j}(r) = \mathbb{P}(X_j \leq r) = r, \qquad 0 \leq r \leq 1.$$

The *empirical distribution function* of the sample $X_1, \ldots, X_m$ is defined by

$$F_m(r) := m^{-1} \sum_{j=1}^{m} 1_{\{X_j \leq r\}}, \qquad 0 \leq r \leq 1,$$

and its properly rescaled version

$$\mathcal{E}_m(r) := \sqrt{m}(F_m(r) - r) = m^{-1/2} \sum_{j=1}^{m} \left(1_{\{X_j \leq r\}} - r\right)$$

is called *empirical process.*

Notice that $\mathbb{E}\,\mathbf{1}_{\{X_j \leq r\}} = r$ and for $r_1 \leq r_2$

$$\mathbb{E}\left(\mathbf{1}_{\{X_j \leq r_1\}}\mathbf{1}_{\{X_j \leq r_2\}}\right) = \mathbb{E}\,\mathbf{1}_{\{X_j \leq r_1\}} = r_1.$$

It follows that

$$\operatorname{cov}\left(\mathbf{1}_{\{X_j \leq r_1\}}, \mathbf{1}_{\{X_j \leq r_2\}}\right) = r_1 - r_1 r_2 = \min\{r_1, r_2\} - r_1 r_2.$$

On the other hand, by independence, for any $i \neq j$ we have equality

$$\operatorname{cov}\left(\mathbf{1}_{\{X_i \leq r_1\}}, \mathbf{1}_{\{X_j \leq r_2\}}\right) = 0.$$

It follows that for any $m \geq 1$

$$\operatorname{cov}\left(\mathcal{E}_m(r_1), \mathcal{E}_m(r_2)\right) = m^{-1} \sum_{j=1}^{m} \operatorname{cov}\left(\mathbf{1}_{\{X_j \leq r_1\}}, \mathbf{1}_{\{X_j \leq r_2\}}\right)$$

$$= \min\{r_1, r_2\} - r_1 r_2$$

coincides with covariance function of Brownian bridge (6.7).

Proposition 11.6. *It is true that, as $m \to \infty$,*

$$\mathcal{E}_m \xrightarrow{f.d.d.} \overset{o}{W},$$

where $\overset{o}{W}$ is a Brownian bridge.

Proof. We have to check that for any $n \geq 1$ and any set of levels $0 \leq r_1 < \cdots < r_n \leq 1$ it is true that $P^{\mathcal{E}_m}_{r_1,\ldots,r_n} \Rightarrow P^{\overset{o}{W}}_{r_1,\ldots,r_n}$. By definition $P^{\mathcal{E}_m}_{r_1,\ldots,r_n}$ is the distribution of the random vector $(\mathcal{E}_m(r_1), \ldots, \mathcal{E}_m(r_n))$ which can be represented as centered and rescaled sum of the i.i.d. vectors $(\mathbf{1}_{\{X_j \leq r_1\}}, \ldots, \mathbf{1}_{\{X_j \leq r_n\}})$. By multivariate Central Limit Theorem (see Subsection 4.4) the distributions of these sums converge to $\mathcal{N}(0, K)$ where K corresponds to the matrix $(\min\{r_i, r_j\} - r_i r_j)_{1 \leq i,j \leq n}$. Since the corresponding vector of Brownian bridge values $(\overset{o}{W}(r_1), \ldots, \overset{o}{W}(r_n))$ also has distribution $\mathcal{N}(0, K)$, we are done. $\square$

Exercise 11.7. Give another proof of Proposition 11.3 based on the multivariate CLT.

Exercise 11.8. Let $X_1, X_2, \ldots$ be i.i.d. random variables with a common distribution function F. Redefine appropriately empirical process $\mathcal{E}_m$ and prove that $\mathcal{E}_m \xrightarrow{f.d.d.} \overset{o}{W}(F(\cdot))$.

Finite-dimensional convergence implies convergence of functionals (11.1) depending only of finite number of values. Unfortunately, the most important functionals do not belong to this class. For example, for a process $X(t), t \in [0,1]$, the functionals $g_1(X) := \max_{0 \leq t \leq 1} X(t)$ and $g_2(X) := \int_0^1 X(t)dt$ depend on the entire uncountable set of process values. Hence $X_m \overset{\text{f.d.d.}}{\longrightarrow} X$ does not imply in general the desirable convergence $g_1(X_m) \Rightarrow g_1(X)$ or $g_2(X_m) \Rightarrow g_2(X)$. This is why stronger types of convergence are needed. One of them is considered in the next subsection.

11.2   *Weak convergence*

We are going to introduce now another type of convergence that opens access to limit theorems for a larger class of functionals than convergence in finite-dimensional distributions. The main idea is to extend the notion of weak convergence (handled earlier for random variables in Section 1 and for random vectors in Section 4) to a more abstract setting of random elements and to represent trajectories of random processes as random elements of appropriate functional space.

What follows is only a crash course giving basic ideas and adjusted to our narrow aims. The books [11, 107] present an excellent introduction to weak convergence theory, see also [83, 105].

11.2.1   *Metric spaces: reminder*

A pair $(\mathcal{X}, \rho)$ of a set $\mathcal{X}$ and a function $\rho : \mathcal{X}^2 \to \mathbb{R}_+$ is called a *metric space* if

- $\rho(x,y) = 0$ iff $x = y$ for all $x, y \in \mathcal{X}$;
- $\rho(x,y) = \rho(y,x)$ for all $x, y \in \mathcal{X}$;
- $\rho(x,z) \leq \rho(x,y) + \rho(y,z)$ for all $x, y, z \in \mathcal{X}$.

The function ρ is called *distance* or *metric*. A set

$$B(x,r) := \{y \in \mathcal{X} : \rho(x,y) < r\}$$

is called open ball of radius r centered at $x \in \mathcal{X}$. A set $A \subset \mathcal{X}$ is called *open* if for any $x \in A$ there exists $r > 0$ such that $B(x,r) \subset A$. A set $A \subset \mathcal{X}$ is called *closed* if its complement $\mathcal{X} \backslash A$ is open.

We say that a point $x \in \mathcal{X}$ is a limit of a sequence $(x_m)_{m \geq 1}$ and write $x_m \to x$ or $x = \lim_m x_m$ iff $\lim_m \rho(x_m, x) = 0$.

A point $x \in \mathcal{X}$ belongs to the *boundary* ∂A of a set $A \subset \mathcal{X}$ if there exist two sequences $x_m \to x$ and $x'_m \to x$ such that $x_m \in A, x'_m \notin A$.

A function $f : \mathcal{X} \to \mathbb{R}$ is called continuous if $x_m \to x$ implies $f(x_m) \to f(x)$.

The minimal sigma-field containing all open sets is called *Borel sigma-field* on $\mathcal{X}$. It is denoted by $\mathcal{B}_{\mathcal{X}}$.

Example 11.9. The space $\mathcal{X} := \mathbb{C}[0,1]$ of all real-valued continuous functions on the interval $[0,1]$ equipped with the *uniform distance*

$$\rho_{\mathbb{C}}(x,y) := \max_{0 \le t \le 1} |x(t) - y(t)|$$

is a metric space.

In many important cases we have to deal with processes with jumps so that spaces like $\mathbb{C}[0,1]$ are inappropriate. A possible remedy is provided by the following construction.

A function $x : [a,b] \to \mathbb{R}$ is called *cadlag*[1] if it is right continuous, i.e.

$$x(t) = \lim_{s \searrow t} x(s), \qquad a \le t < b,$$

and there exist finite left limits

$$x_-(t) := \lim_{s \nearrow t} x(s), \qquad a < t \le b.$$

The space $\mathbb{D}[0,1]$ of all cadlag functions on $[0,1]$ is called *Skorokhod space*. A.V. Skorokhod who introduced this space in [94, 95] suggested several ways to build a distance on $\mathbb{D}[0,1]$ and thus consider it as a metric space. We will use only two distances from his set, so called J-distance and M-distance. In original terminology they are called J_1- and M_1-distance, respectively. In the literature, J-distance is used by far more often than M-distance, so that it is plainly referred to as a "Skorokhod distance". However, for our purposes we will rather need M-distance. We refer to [107], Chapter 12 for detailed comparison of two distances.

Skorokhod J-distance between two cadlag functions is defined by

$$\rho_{\mathbb{D},J}(x,y) := \inf_{u \in \mathcal{U}_J} \left[\sup_{0 \le t \le 1} |x(t) - y(u(t))| + \sup_{0 \le t \le 1} |t - u(t)| \right], \qquad (11.2)$$

where $\mathcal{U}_J$ is the class of all increasing continuous mappings (variable changes) from $[0,1]$ onto itself.

---

[1] Abbreviation *cadlag* comes from French "continue à droite, limite à gauche" which means "right continuous, with left limits".

If $\rho_{_{\mathbb{D},J}}(x,y) \leq \varepsilon$ and the function $x(\cdot)$ has a jump $\delta := x(t) - x_-(t)$ such that $|\delta| > \varepsilon$ at a point t, then y has a jump $y(t') - y_-(t') = \delta \pm \varepsilon$ at some neighboring point $t' = t \pm \varepsilon$.

Exercise 11.10. Check that (11.2) is a distance.

Exercise 11.11. Check that $\mathbb{C}[0,1]$ is a J-closed subset of $\mathbb{D}[0,1]$. In other words, if a sequence of functions $(x_n) \subset \mathbb{C}[0,1] \subset \mathbb{D}[0,1]$, and $\rho_{_{\mathbb{D},J}}(x_n, x) \to 0$, then $x \in \mathbb{C}[0,1]$.

Skorokhod M-distance between two cadlag functions reflects the closeness between their graphs. Let us recall the corresponding definitions. For two real numbers a, b let denote a *segment*
$$[a,b] := \{\alpha a + (1-\alpha)b; \ 0 \leq \alpha \leq 1\}.$$
The *graph* of a cadlag function x is defined by
$$\Gamma_x := \{(z,t) : 0 \leq t \leq 1, z \in [x_-(t), x(t)]\}.$$
Notice that Γ_x is becomes a totally ordered set, if we let
$$(z_1, t_1) \leq (z_2, t_2) \text{ iff } \begin{cases} t_1 \leq t_2, \text{ or} \\ t_1 = t_2, \ |z_1 - x_-(t_1)| \leq |z_2 - x_-(t_1)|. \end{cases} \tag{11.3}$$
We call a mapping u from $[0,1]$ onto Γ_x a *parametric representation* of Γ_x if it is non-decreasing with respect to the order (11.3). Let $\mathcal{U}_{M,x}$ be the class of all parametric representations of Γ_x.

Finally, Skorokhod M-distance is defined by
$$\rho_{_{\mathbb{D},M}}(x,y) := \inf_{\substack{u_1 \in \mathcal{U}_{M,x} \\ u_2 \in \mathcal{U}_{M,y}}} \sup_{0 \leq t \leq 1} \|u_1(t) - u_2(t)\|_\infty,$$
where $\|(z,t)\|_\infty := \max\{|z|, |t|\}$ for $(z,t) \in \mathbb{R}^2$.

J-convergence is stronger than M-convergence in a sense that if $\rho_{_{\mathbb{D},J}}(x_n, x) \to 0$, then $\rho_{_{\mathbb{D},M}}(x_n, x) \to 0$, but the converse is not true in general, as one can see from the following example.

Exercise 11.12. Let $x(t) = 1_{\{[1/2,1]\}}(t)$ be a one-jump cadlag function on $[0,1]$. For $n > 2$, let $x_n(t) := \frac{1}{2} 1_{\{[1/2, 1/2+1/n)\}}(t) + 1_{\{[1/2+1/n, 1]\}}(t)$ be two-jump cadlag functions, and let piecewise linear continuous functions y_n be given by
$$y_n(t) := \begin{cases} 0 & 0 \leq t \leq 1/2; \\ n(t - 1/2), & 1/2 \leq t \leq 1/2 + 1/n, \\ 1, & 1/2 + 1/n \leq t \leq 1. \end{cases}$$
Prove that $\rho_{_{\mathbb{D},M}}(x_n, x) \to 0$, $\rho_{_{\mathbb{D},M}}(y_n, x) \to 0$ but $\rho_{_{\mathbb{D},J}}(x_n, x) \geq \frac{1}{2}$ and $\rho_{_{\mathbb{D},J}}(y_n, x) \geq \frac{1}{2}$ for any $n > 2$.

We see from this example that $\mathbb{C}[0,1]$ is not an M-closed subset of $\mathbb{D}[0,1]$.

If $\tilde{X} : \Omega \to \mathbb{C}[0,1]$ or $\tilde{X} : \Omega \to \mathbb{D}[0,1]$ is a random element of the corresponding function space, then $\tilde{X}(t), 0 \leq t \leq 1$, is a random process. It is important for us to know when a given random process $X(t), t \in [0,1]$, can be represented in this form at least on the level of finite-dimensional distributions. In other words, we ask whether there exists a random element $\tilde{X}$ in $\mathcal{X} = \mathbb{C}[0,1]$ or in $\mathcal{X} = \mathbb{D}[0,1]$ such that the processes $X(t), t \in [0,1]$, and $\tilde{X}(t), 0 \leq t \leq 1$, have the same finite-dimensional distributions. The answer, in terms of increments' moments, will be given in the two following propositions.

Theorem 11.13 (Kolmogorov continuity criterion).
Consider a random process $X(t), t \in [0,1]$. Assume that one of the two conditions holds:

a) For some $a > 0, C > 0$ and $b > 1$ it is true that
$$\mathbb{E}\,|X(t_1) - X(t_2)|^a \leq C|t_2 - t_1|^b \qquad \text{for all } 0 \leq t_1, t_2 \leq 1.$$

b) X is a centered Gaussian process and for some $C > 0$ and $b > 1$ it is true that
$$\mathbb{E}\,|X(t_1) - X(t_2)|^2 \leq C\big|\ln|t_2 - t_1|\big|^{-b} \qquad \text{for all } 0 \leq t_1, t_2 \leq 1. \quad (11.4)$$
Then there exists a random element $\tilde{X} \in \mathbb{C}[0,1]$ such that the processes $X(t), t \in [0,1]$, and $\tilde{X}(t), 0 \leq t \leq 1$, have the same finite-dimensional distributions.

For the proof of assertions a) and b) we refer to the books [11], Theorem 12.4, resp. [60], Section 15.

Example 11.14. Let B^H be a fractional Brownian motion. Then by stationarity of increments and self-similarity we have
$$\mathbb{E}\,|B^H(t_1) - B^H(t_2)|^a = C_a|t_2 - t_1|^{Ha}$$
where $C_a := \mathbb{E}\,|Y|^a$ for a standard normal variable Y. Therefore, (11.4) holds whenever a is chosen so large that $Ha > 1$. In particular, Theorem 11.13 applies to Wiener process, since the latter is a particular case of fBm. Moreover, Theorem 11.13 also applies to a Brownian bridge since
$$\begin{aligned}
\mathbb{E}\,(\overset{\circ}{W}(t_1) - \overset{\circ}{W}(t_2))^2 &= \mathbb{E}\,\overset{\circ}{W}(t_1)^2 + \mathbb{E}\,\overset{\circ}{W}(t_2)^2 - 2\,\mathbb{E}\,(\overset{\circ}{W}(t_1)\,\overset{\circ}{W}(t_2)) \\
&= (t_1 - t_1^2) + (t_2 - t_2^2) - 2\min\{t_1, t_2\} + 2t_1 t_2 \\
&= (t_1 + t_2 - 2\min\{t_1, t_2\}) - (t_1 - t_2)^2 \\
&= |t_1 - t_2| - (t_1 - t_2)^2 \leq |t_1 - t_2|,
\end{aligned}$$

whence

$$\mathbb{E}\,|\overset{\circ}{W}(t_1)-\overset{\circ}{W}(t_2)|^a \le C_a|t_2 - t_1|^{a/2}.$$

Theorem 11.15. *Let $X(t), t \in [0,1]$, be a random process. Assume that for some $a > 0, C > 0$ and $b > 1$ it is true that*

$$\mathbb{E}\,|X(t_1) - X(t)|^a|X(t_2) - X(t)|^a \le C|t_2 - t_1|^b \qquad (11.5)$$

for all $0 \le t_1 \le t \le t_2 \le 1$.

Then there exists a random element $\tilde{X} \in \mathbb{D}[0,1]$ such that $X(t), t \in [0,1]$, and $\tilde{X}(t), 0 \le t \le 1$, have the same finite-dimensional distributions.

For the proof we refer to [11], Theorem 15.7.

Example 11.16. We show now that any Lévy process may be represented as a random element of $\mathbb{D}[0,1]$. Take first any Lévy process X with zero mean and finite variance. Then for all $0 \le t_1 \le t \le t_2 \le 1$

$$\mathbb{E}\left[|X(t_1) - X(t)|^2|X(t_2) - X(t)|^2\right] = \mathbb{E}\,|X(t_1) - X(t)|^2\,\mathbb{E}\,|X(t_2) - X(t)|^2$$
$$= C(t - t_1) \cdot C(t_2 - t) \le C^2(t_2 - t_1)^2,$$

where $C := \mathbb{E}\,X(1)^2$. Hence, (11.5) holds and Theorem 11.15 applies. Next, if X is a compound Poisson process defined in Subsection 9.2, then the definition (9.4) itself provides a representation of X as a random element of Skorokhod space (note that this element is a random step function with Poissonian number of steps). Finally, arbitrary Lévy process X may be represented as a sum of three independent components: a centered Lévy process with finite variance, a compound Poisson process, and a deterministic linear function. This representation is achieved by splitting the triplet (a, σ^2, ν) of X into the corresponding parts $(0, \sigma^2, \nu\mathbf{1}_{\{[-1,1]\}})$, $(0, 0, \nu\mathbf{1}_{\{\mathbb{R}\setminus[-1,1]\}})$, and $(a, 0, 0)$. The sum of independent random elements corresponding to the parts of representation provides a required representation for X.

11.2.2 Weak convergence

Let $(\Omega, \mathcal{A}, \mathbb{P})$ be a probability space and $\mathcal{X}$ a metric space. A measurable mapping $X : (\Omega, \mathcal{A}) \mapsto (\mathcal{X}, \mathcal{B}_\mathcal{X})$ is called an $\mathcal{X}$-valued *random element*. Here "measurable" means that for any $B \in \mathcal{B}_\mathcal{X}$ we have

$$X^{-1}(B) := \{\omega \in \Omega : X(\omega) \in B\} \in \mathcal{A}.$$

The *distribution* of X is a measure on $(\mathcal{X}, \mathcal{B}_\mathcal{X})$ defined by

$$P_X(B) := \mathbb{P}(X \in B), \qquad B \in \mathcal{B}_\mathcal{X}.$$

Exactly as in Theorem 4.2 we have the following equivalent definitions (see [11], Theorem 2.1).

Theorem 11.17. *Let* $(X_k)_{k\geq 1}$ *and* X *be some* $\mathcal{X}$-*valued random elements. Then the following assertions are equivalent.*

a) For any closed set $B \subset \mathcal{X}$ *it is true that*

$$\limsup_{k\to\infty} \mathbb{P}\{X_k \in B\} \leq \mathbb{P}\{X \in B\}.$$

b) For any open set $V \subset \mathcal{X}$ *it is true that*

$$\liminf_{k\to\infty} \mathbb{P}\{X_k \in V\} \geq \mathbb{P}\{X \in V\}.$$

c) For any measurable set $A \subset \mathcal{X}$, *satisfying* regularity condition *with respect to* X, *i.e.* $\mathbb{P}(X \in \partial A) = 0$, *it is true that*

$$\lim_{k\to\infty} \mathbb{P}\{X_k \in A\} = \mathbb{P}\{X \in A\}.$$

d) For any bounded continuous function $f : \mathcal{X} \mapsto \mathbb{R}$ *it is true that*

$$\lim_{k\to\infty} \mathbb{E}\, f(X_k) = \mathbb{E}\, f(X).$$

If any of properties a) – d) holds, we say that the sequence X_k *converges in distribution*, or *converges in law*, or *converges weakly* to X and write $X_k \Rightarrow X$.

Corollary 11.18. *Let* $X_k \Rightarrow X$ *in* $\mathcal{X}$ *and let* $g : \mathcal{X} \to \mathbb{R}$ *be a continuous function. Then* $g(X_k) \Rightarrow g(X)$.

Proof. Take any bounded continuous function $f : \mathbb{R} \mapsto \mathbb{R}$. The function $f(g(\cdot))$ is bounded and continuous on $\mathcal{X}$. Hence, by Theorem 11.17 d) we have

$$\lim_{k\to\infty} \mathbb{E}\, f(g(X_k)) = \mathbb{E}\, f(g(X)).$$

Now from Theorem 1.21 e) it follows that $g(X_k) \Rightarrow g(X)$. $\qquad\qquad\square$

Remark 11.19. The assertion of Corollary 11.18 remains true if we only assume P_X-almost everywhere continuity of $g(\cdot)$, i.e.

$$\mathbb{P}\{\omega : g(\cdot) \text{ is continuous at the point } X(\omega)\} = 1.$$

Now we present specific sufficient conditions of weak convergence in spaces $\mathbb{C}[0,1]$ and $\mathbb{D}[0,1]$. Roughly speaking, weak convergence comes from f.d.d.-convergence with additional assumption of *uniform* "existence conditions" from Theorems 11.13 and 11.15. More precisely, we have the following.

Theorem 11.20. *Let a sequence (X_k) and X be random elements of $\mathbb{C}[0,1]$. Assume that*

$$X_k \xrightarrow{f.d.d.} X$$

and that for any $\varepsilon, \eta > 0$ there exist $t = t(\varepsilon, \eta)$ and $k_0 = k_0(\varepsilon, \eta)$ such that it is true

$$\sup_{k \geq k_0} \sup_{0 \leq \theta \leq 1} \mathbb{P}\left(\sup_{\substack{\theta \leq \tau \leq \theta + t \\ \tau \leq 1}} |X_k(\tau) - X_k(\theta)| \geq \varepsilon \right) \leq \eta t. \qquad (11.6)$$

Then $X_k \Rightarrow X$ in $\mathbb{C}[0,1]$.

Assumption (11.6) holds if for some $a > 0, C > 0$ and $b > 1$ it is true that for all $k \geq 1$

$$\mathbb{E}|X_k(t_1) - X_k(t_2)|^a \leq C|t_2 - t_1|^b \qquad \text{for all } 0 \leq t_1 \leq t_2 \leq 1. \qquad (11.7)$$

For the proof we refer to [11], Theorems 8.1, 8.3, and 12.3.

Now we discuss the weak convergence in $\mathbb{D}[0,1]$. Since J-convergence and M-convergence are not equivalent, the classes of open and closed sets related to the distances $\rho_{\mathbb{D},J}$ and $\rho_{\mathbb{D},M}$ do not coincide. Therefore, we have to distinguish weak convergence in the metric space $(\mathbb{D}[0,1], \rho_{\mathbb{D},M})$ and that in $(\mathbb{D}[0,1], \rho_{\mathbb{D},M})$.

Theorem 11.21. *Let a sequence (X_k) and X be random elements of $\mathbb{D}[0,1]$. Assume that*

$$X_k \xrightarrow{f.d.d.} X$$

and that for some $a > 0, C > 0$ and $b > 1$ it is true that for all $k \geq 1$ and for all $0 \leq t_1 \leq t \leq t_2 \leq 1$

$$\mathbb{P}\left(\min\left\{|X_k(t) - X_k(t_1)|, |X_k(t) - X_k(t_2)|\right\} \geq \varepsilon\right) \leq C\varepsilon^{-\nu}|t_2 - t_1|^b. \quad (11.8)$$

Then $X_k \Rightarrow X$ in $(\mathbb{D}[0,1], \rho_{\mathbb{D},J})$.

For the proof we refer to [11], Theorems 15.1 and 15.6.

Notice that (11.8) holds, if for all $0 \leq t_1 \leq t \leq t_2 \leq 1$ the following moment condition is satisfied,

$$\mathbb{E}|X_k(t_1) - X_k(t)|^a |X_k(t_2) - X_k(t)|^a \leq C|t_2 - t_1|^b. \qquad (11.9)$$

Theorem 11.22. *Let a sequence* (X_k) *and* X *be random elements of* $\mathbb{D}[0,1]$. *Assume that*

$$X_k \xrightarrow{f.d.d.} X$$

and that for some $C > 0$ *and* $b > 1$, *for all* $0 \le t_1 \le t \le t_2 \le 1$ *and for all* $k \ge 1$ *it is true that*

$$\mathbb{P}\left(\min_{x \in [X_k(t_1), X_k(t_2)]} |X_k(t) - x| \ge \varepsilon\right) \le C\varepsilon^{-\nu}|t_2 - t_1|^b. \tag{11.10}$$

Then $X_k \Rightarrow X$ *in* $(\mathbb{D}[0,1], \rho_{\mathbb{D},M})$.

The proof of this theorem follows by combining a results by Avram and Taqqu [4], Theorem 1, and [5], Proposition 2, the latter coming from [94].

11.2.3 *Basic examples of weak convergence*

Our first aim is to obtain the weak convergence in the setting of invariance principle for random walks stated in Proposition 11.3. So let again $X_1, X_2, \ldots$ be a sequence of independent identically distributed random variables. Let the random walk be defined by $S_0 := 0$ and

$$S_n := \sum_{j=1}^{n} X_j, \qquad n \ge 1,$$

and let

$$S(t) := S_n, \qquad n \le t < n+1, \ n \ge 0.$$

The following two versions of invariance principles are due to M. Donsker [28]. See also [11], Theorems 16.1 and 10.1.

Theorem 11.23 (Donsker invariance principle in $\mathbb{D}$**).** *Assume that the steps of a random walk satisfy* $\mathbb{E}\, X_j = 0$ *and* $0 < \mathbb{E}\, X_j^2 := \sigma^2 < \infty$. *Let*

$$Z_m(t) := \frac{S(mt)}{\sigma\sqrt{m}}, \qquad 0 \le t \le 1.$$

Then, as $m \to \infty$,

$$Z_m \Rightarrow W \text{ in } (\mathbb{D}[0,1], \rho_{\mathbb{D},J}),$$

where W *is a Wiener process.*

Proof.   Since we have finite-dimensional convergence by Proposition 11.3, by Theorem 11.21 it remains to verify the moment inequality for increments (11.9) with appropriate a, b and C.

Notice that by independence of steps for any $k_1 \leq k \leq k_2$ we have

$$
\mathbb{E}\left[(S_k - S_{k_1})^2 (S_{k_2} - S_k)^2\right] = \mathbb{E}\left[\left(\sum_{j=k_1+1}^{k} X_j\right)^2 \left(\sum_{j=k+1}^{k_2} X_j\right)^2\right]
$$

$$
= \mathbb{E}\left(\sum_{j=k_1+1}^{k} X_j\right)^2 \mathbb{E}\left(\sum_{j=k+1}^{k_2} X_j\right)^2
$$

$$
= \sigma^2 (k - k_1)\, \sigma^2 (k_2 - k) \leq \sigma^4 (k_2 - k_1)^2.
$$

It follows that for any $t_1 \leq t \leq t_2$ we have

$$
\mathbb{E}\left[(Z_m(t) - Z_m(t_1))^2 (Z_m(t_2) - Z_m(t)\right]
$$

$$
= m^{-2}\mathbb{E}\left[(S_{[mt]} - S_{[mt_1]})^2 (S_{[mt_2]} - S_{[mt]})^2\right]
$$

$$
\leq m^{-2}\sigma^4 \left([mt_2] - [mt_1]\right)^2
$$

$$
\leq m^{-2}\sigma^4 \left(mt_2 - mt_1 + 1\right)^2
$$

$$
= \sigma^4 \left(t_2 - t_1 + \frac{1}{m}\right)^2.
$$

If $t_2 - t_1 \geq \frac{1}{m}$, we obtain

$$
\mathbb{E}\left[(Z_m(t) - Z_m(t_1))^2 (Z_m(t_2) - Z_m(t))^2\right] \leq 4\sigma^4 \left(t_2 - t_1\right)^2. \qquad (11.11)
$$

On the other hand, if $t_2 - t_1 < \frac{1}{m}$, we have

$$
(Z_m(t) - Z_m(t_1))(Z_m(t_2) - Z_m(t)) = 0,
$$

because Z_m is a piecewise constant function with step widths $\frac{1}{m}$. Therefore, (11.11) is true for all $t_1 \leq t \leq t_2$, thus (11.9) holds with $a = b = 2$ uniformly in m and Theorem 11.21 applies. $\square$

Now we wish to show a similar result in the space of continuous functions $\mathbb{C}[0, 1]$. However, we have to adjust our definition because the random walk $S(\cdot)$ has jumps. Therefore, we introduce a linearly interpolated version by

$$
\tilde{S}(t) := S_n + (t-n)(S_{n+1} - S_n) = S_n + (t-n)X_{n+1}, \qquad n \leq t < n+1,\ n \geq 0.
$$

Theorem 11.24 (Donsker invariance principle in $\mathbb{C}$). *Assume   that the steps of a random walk satisfy* $\mathbb{E}\, X_j = 0$ *and* $0 < \mathbb{E}\, X_j^2 := \sigma^2 < \infty$. *Let*

$$
\tilde{Z}_m(t) := \frac{\tilde{S}(mt)}{\sigma\sqrt{m}}, \qquad 0 \leq t \leq 1.
$$

Then, as $m \to \infty$,

$$\widetilde{Z}_m \Rightarrow W \ \text{in} \ \mathbb{C}[0,1],$$

where W is a Wiener process.

Proof. At first, we prove that $Z_m(\cdot)$ and $\widetilde{Z}_m(\cdot)$ are uniformly close with high probability. Notice that,

$$\max_{0 \le t \le 1} |\widetilde{Z}_m(t) - Z_m(t)| = \frac{\max_{1 \le j \le m} |X_j|}{\sigma \sqrt{m}} \ .$$

Therefore, for each $\varepsilon > 0$ we have

$$\mathbb{P}\left(\max_{0 \le t \le 1} |\widetilde{Z}_m(t) - Z_m(t)| \ge \varepsilon \right)$$

$$= \mathbb{P}\left(\max_{1 \le j \le m} |X_j| \ge \sigma \varepsilon \sqrt{m} \right)$$

$$\le \sum_{j=1}^m \mathbb{P}\left(|X_j| \ge \sigma \varepsilon \sqrt{m} \right)$$

$$= m \, \mathbb{P}\left(|X_1| \ge \sigma \varepsilon \sqrt{m} \right)$$

$$\le m \, \mathbb{E}\left(\frac{X_1^2}{\sigma^2 \varepsilon^2 m} \mathbf{1}_{\{|X_1| \ge \varepsilon \sqrt{m}\}} \right)$$

$$= \sigma^{-2} \varepsilon^{-2} \mathbb{E}\left(X_1^2 \mathbf{1}_{\{|X_1| \ge \varepsilon \sqrt{m}\}} \right) \to 0, \tag{11.12}$$

as $m \to \infty$.

Let B be a closed set in $\mathbb{C}[0,1]$. Then B also is a closed set in $(\mathbb{D}[0,1], \rho_{_{\mathbb{D},J}})$. Indeed, let a sequence (x_n) in B converge to an $x \in \mathbb{D}[0,1]$, which means $\rho_{_{\mathbb{D},J}}(x_n, x) \to 0$. In other words, there exists a sequence of continuous variable changes u_n (uniformly approaching identity function) such that $\rho_{_{\mathbb{C}}}(x_n(u_n(\cdot)), x) \to 0$. Hence, x is a continuous function, as a uniform limit of continuous functions. We have

$$\rho_{_{\mathbb{C}}}(x_n, x) = \rho_{_{\mathbb{C}}}(x_n(u_n(\cdot)), x(u_n(\cdot)))$$

$$\le \rho_{_{\mathbb{C}}}(x_n(u_n(\cdot)), x) + \rho_{_{\mathbb{C}}}(x, x(u_n(\cdot))) \to 0.$$

Since B is closed in $\mathbb{C}[0,1]$, this yields $x \in B$. Hence, B is closed in $(\mathbb{D}[0,1], \rho_{_{\mathbb{D},J}})$.

Next, for any $\varepsilon > 0$ let

$$B^\varepsilon := \{ y \in \mathbb{D}[0,1] : \inf_{x \in B} \rho_{_{\mathbb{D},J}}(x, y) \le \varepsilon \}$$

be the closed ε-enlargement of B in $\mathbb{D}[0,1]$. For each m and $\varepsilon > 0$ we have

$$\mathbb{P}(\widetilde{Z}_m \in B) = \mathbb{P}(\widetilde{Z}_m \in B; \rho_{\mathbb{D},J}(\widetilde{Z}_m, Z_m) \leq \varepsilon) + \mathbb{P}(\widetilde{Z}_m \in B; \rho_{\mathbb{D},J}(\widetilde{Z}_m, Z_m) > \varepsilon)$$
$$\leq \mathbb{P}(Z_m \in B^{\varepsilon}) + \mathbb{P}(\rho_{\mathbb{D},J}(\widetilde{Z}_m, Z_m) > \varepsilon)$$
$$\leq \mathbb{P}(Z_m \in B^{\varepsilon}) + \mathbb{P}(\rho_{c}(\widetilde{Z}_m, Z_m) > \varepsilon)$$
$$= \mathbb{P}(Z_m \in B^{\varepsilon}) + \mathbb{P}\left(\max_{0 \leq t \leq 1} |\widetilde{Z}_m(t) - Z_m(t)| > \varepsilon\right).$$

By taking the limits, using Theorem 11.23 and (11.12) we obtain

$$\limsup_{m \to \infty} \mathbb{P}(\widetilde{Z}_m \in B) \leq \limsup_{m \to \infty} \mathbb{P}(Z_m \in B^{\varepsilon})$$
$$+ \lim_{m \to \infty} \mathbb{P}\left(\max_{0 \leq t \leq 1} |\widetilde{Z}_m(t) - Z_m(t)| > \varepsilon\right)$$
$$\leq \mathbb{P}(W \in B^{\varepsilon}).$$

By letting $\varepsilon \searrow 0$, we obtain (using that B is closed in $\mathbb{D}[0,1]$)

$$\limsup_{m \to \infty} \mathbb{P}(\widetilde{Z}_m \in B) \leq \lim_{\varepsilon \searrow 0} \mathbb{P}(W \in B^{\varepsilon}) = \mathbb{P}\left(W \in \cap_{\varepsilon > 0} B^{\varepsilon}\right) = \mathbb{P}(W \in B),$$

as required in the first definition of weak convergence, cf. Theorem 11.17 a).   □

Remark 11.25. There exist many extensions of invariance principle, see e.g. [84], where sums of non-identically distributed variables and schemes with stable limits are considered.

Remark 11.26. There exits an important alternative approach to invariance principle based on the joint construction of close copies of $S(\cdot)$ and $W(\cdot)$ on a common probability space. See more about those results in the references [23, 30, 52, 53, 64, 88, 92, 96, 115, 116].

Our next aim is to obtain the weak convergence for empirical processes considered in Proposition 11.6. So let $X_1, X_2, \ldots$ be i.i.d. random variables uniformly distributed on the interval $[0,1]$. Recall that the corresponding empirical process is defined by

$$\mathcal{E}_m(r) = m^{-1/2} \sum_{j=1}^{m} \left(\mathbf{1}_{\{X_j \leq r\}} - r\right).$$

Theorem 11.27. *It is true that, as $m \to \infty$,*

$$\mathcal{E}_m \Rightarrow \overset{\circ}{W} \quad in \ (\mathbb{D}[0,1], \rho_{\mathbb{D},J}),$$

where $\overset{\circ}{W}$ is a Brownian bridge.

Proof.   Since we have finite-dimensional convergence by Proposition 11.6, by Theorem 11.21 it remains to verify the moment inequality for increments (11.9) with appropriate a, b and C. Fix some $t_1 \leq t \leq t_2$ and let $\ell_1 := t - t_1$ and $\ell_2 := t_2 - t$ be the lengths of the corresponding intervals. Then

$$\mathbb{E}\,|\mathcal{E}_m(t_1) - \mathcal{E}_m(t)|^2 |\mathcal{E}_m(t_2) - \mathcal{E}_m(t)|^2$$

$$= m^{-2}\mathbb{E}\left(\sum_{j=1}^{m}\left(\mathbf{1}_{\{X_j \in (t_1, t]\}} - \ell_1\right)\right)^2 \left(\sum_{j=1}^{m}\left(\mathbf{1}_{\{X_j \in (t, t_2]\}} - \ell_2\right)\right)^2$$

$$:= m^{-2}\mathbb{E}\left(\sum_{j=1}^{m} u_j\right)^2 \left(\sum_{j=1}^{m} v_j\right)^2$$

$$= m^{-2}\sum_{1 \leq j_1, j_2 \leq m}\,\sum_{1 \leq i_1, i_2 \leq m}\mathbb{E}\left(u_{j_1} u_{j_2} v_{i_1} v_{i_2}\right).$$

Since all variables here are centered and the variables with different subscripts are independent, many expectations vanish and we have

$$\mathbb{E}\,|\mathcal{E}_m(t_1) - \mathcal{E}_m(t)|^2 |\mathcal{E}_m(t_2) - \mathcal{E}_m(t)|^2 = m^{-2}\left(S_1 + S_2 + S_3\right),$$

where

$$S_1 = \sum_{1 \leq j \leq m}\mathbb{E}\left(u_j^2 v_j^2\right) = m\,\mathbb{E}\left(u_1^2 v_1^2\right)$$

$$= m\left[\ell_1(1 - \ell_1)^2\ell_2^2 + \ell_2\ell_1^2(1 - \ell_2)^2 + (1 - \ell_1 - \ell_2)\ell_1^2\ell_2^2\right] \leq 3m\ell_1\ell_2;$$

$$S_2 = 2\sum_{1 \leq j < i \leq m}\mathbb{E}\left(u_j^2 v_i^2\right) = m(m-1)\mathbb{E}\left(u_1^2 v_2^2\right) \leq m^2\mathbb{E}\left(u_1^2\right)\mathbb{E}\left(v_2^2\right)$$

$$= m^2\ell_1(1 - \ell_1)\,\ell_2(1 - \ell_2) \leq m^2\ell_1\ell_2;$$

$$S_3 = 4\sum_{1 \leq j < i \leq m}\mathbb{E}\left(u_j v_j u_i v_i\right) = 2m(m-1)\mathbb{E}\left(u_1 v_1 u_2 v_2\right) \leq 2m^2\left[\mathbb{E}\left(u_1 v_1\right)\right]^2$$

$$= 2m^2\left(\ell_1(1 - \ell_1)(-\ell_2) + \ell_2(1 - \ell_2)(-\ell_1)(1 - \ell_1 - \ell_2)(-\ell_1)(-\ell_2)\right)^2$$

$$= 2m^2(-\ell_1\ell_2)^2 = 2m^2\ell_1^2\ell_2^2 \leq 2m^2\ell_1\ell_2.$$

We arrive at

$$\mathbb{E}\,|\mathcal{E}_m(t_1) - \mathcal{E}_m(t)|^2 |\mathcal{E}_m(t_2) - \mathcal{E}_m(t)|^2 \leq 6\ell_1\ell_2 \leq 6(\ell_1 + \ell_2)^2 = (t_2 - t_1)^2,$$

and (11.9) follows with $a = b = 2$. $\square$

Further interesting examples of convergence in $(\mathbb{D}[0, 1], \rho_{\mathbb{D}, M})$ are given in the next chapter.

Chapter 3

A Playground: Teletraffic Models

In this chapter, we illustrate the general theory of random processes by considering a simple and intuitively trivial model of a service system. Quite surprisingly, a minor tuning of few system's parameters leads to different system's workload regimes – Wiener process, fractional Brownian motion, stable Lévy process, as well as to some less commonly known ones, called "Telecom processes". Therefore, we may highlight the different aspects of the theory with very few material at hand.

Our subject comes from the remarkable work by I. Kaj and M.S. Taqqu [49] who handled the limit behavior of "teletraffic systems" by using the language of integral representations as a unifying technique. Their article brilliantly represents a wave of interest to the subject, see e.g. [8, 10, 13, 14, 24, 25, 34, 36, 39, 48, 54, 57, 69–71, 80, 82, 87, 101], [108–111], and the surveys with further references [44–47], to mention just a few. Simplicity of the dependence mechanism used in the model enables to get a clear understanding both of long range dependence in one case, and independent increments, in other cases.

However, there is not too much stuff specific for teletraffic in the model of a service system used in [49]. Not surprisingly, this approach, as a true good mathematics, found applications in other domains quite distant from telecommunication, e.g. in medicine (modelling of internal bone structure).

In what concerns the mathematical problem setting and results, we basically follow [49]. However, for teaching and learning purposes, our proofs and the terminology used throughout the exposition are often different from the original ones.

We recommend to the reader to explore the model description and the chart of limit theorems in the beginning of the chapter, then to study carefully the limit theorems' statements. As for the theorems' proofs, which are rather technical, their reading is optional. One might read a sample and try to prove the remaining by himself using the tools from previous chapters.

12 A Model of Service System

The work of our system represents a collection of *service processes* or *sessions*, if one wants to use telecommunication terminology. Every process starts at some time s, lasts u units of time, and occupies r *resource* units (synonyms for resource are *reward, transmission rate* etc). The amount of occupied resources r remains constant during every service process, from the start time s till its end time $s + u$. We say that a service process is *active* at time t if $s \leq t \leq s + u$.

Many service processes may be handled in the system simultaneously, and there is no limit on the service capacity of the system, which is of course a drastic simplification with respect to reality.[1]

Therefore, the system's *instant workload*, or the amount of occupied resources at time t, is the sum of occupied resources over processes active at time t:

$$W^{\circ}(t) := \sum_j r_j \mathbf{1}_{\{s_j \leq t \leq s_j + u_j\}}.$$

We will be mainly interested in the system's *integral* (or *aggregated*) workload,

$$W^*(t) := \int_0^t W^{\circ}(\tau)d\tau,$$

which somehow describes the total amount of resources used by the system on the time interval $[0,t]$ for handling the ensemble of service processes. Let us stress that although we start workload observation from initial time 0, the system itself is assumed to operate on bilaterally infinite horizon of time $(-\infty, +\infty)$. This means that eventually we have some service processes that started at negative time s and were already active when we started our observation, as well as some processes that start later, during the observation period.

We assume that starting time, the duration, and the required amount of resources are *random variables*. Moreover, it is natural to assume that

---

[1]Limitation of service capacity leads to a deep *Queueing Theory* which can not be considered here, see e.g. [38, 67, 75, 76, 118].

the characteristics of different service processes have the same distribution and that they are independent.

It is also natural to assume that the system works in a stationary regime, e.g. with the same distribution of number of service processes starting at time intervals of equal length.

Moreover, following [49], we will assume that the duration and the number of required resources for a particular service process are independent. This is a much less natural assumption.[2] Many results described below remain valid without it. However, handling the technicalities of dependent case would divert us from focusing on the fundamental principles.

Now we inject more rigor into the model description.

The formal model of the service system is based on Poisson random measures and looks as follows. Let $\mathcal{R} := \{(s, u, r)\} = \mathbb{R} \times \mathbb{R}_+ \times \mathbb{R}_+$. Every point (s, u, r) corresponds to a possible service process with starting time s, duration u, and required resources r.

The system is characterized by the following parameters:

- $\lambda > 0$ – *arrival intensity* of service processes, i.e. the average number of processes starting during any time interval of unit length;
- $F_U(du)$ – the distribution of service duration;
- $F_R(dr)$ – the distribution of amount of required resources.

Define on $\mathcal{R}$ an intensity measure

$$\mu(ds, du, dr) = \lambda ds\, F_U(du)\, F_R(dr).$$

Let N be a corresponding Poisson random measure. One may consider the samples of random measure N (sets of triplets (s, u, r), each triplet corresponding to a service process) as variants (sample paths) of the work for our system.

Notice that the system is stationary in time due to shift-invariance of the Lebesgue measure ds controlling time in intensity measure. The product structure $F_U(du)\, F_R(dr)$ instead of eventual more general joint distribution $F_{UR}(du, dr)$ corresponds to the independence of duration and resource consummation for given service process.

Now we are able to express many characteristics of the system as the corresponding Poisson integrals.

In particular, the instant workload on the system at time t writes as

$$W^\circ(t) = \int_{\mathcal{R}} r \mathbf{1}_{\{s \le t \le s+u\}} dN, \qquad (12.1)$$

---

[2]See the discussion of independence hypothesis in [25].

while the integral workload over the interval $[0, t]$ is

$$W^*(t) = \int_0^t W^\circ(\tau)d\tau = \int_{\mathcal{R}} r \int_0^t \mathbf{1}_{\{s \leq \tau \leq s+u\}} d\tau dN \qquad (12.2)$$

$$= \int_{\mathcal{R}} r \cdot \Big|[s, s+u] \cap [0, t]\Big| dN := \int_{\mathcal{R}} r\ell_t(s, u)dN. \qquad (12.3)$$

Here $|\cdot|$ stands for the length of an interval, and the kernel

$$\ell_t(s, u) := \Big|[s, s+u] \cap [0, t]\Big| \qquad (12.4)$$

that appears here for the first time will be often used in the sequel. There is no problem to write down an exact lengthy formula for $\ell_t(s, u)$ but the graphical presentation is more convenient, see Figures 12.2 and 12.2. We will mostly need the trivial inequalities $\ell_t(s, u) \leq u$ and $\ell_t(s, u) \leq t$.

Notice that $W^\circ(\cdot)$ is a stationary process and its integral $W^*(\cdot)$ is a process with stationary increments, cf. Exercise 5.1.

12.1 Main assumptions on the service time and resource consummation

Of course, in order to obtain any reasonable properties of the system, some assumptions should be made about the distributions F_R and F_U. For writing convenience, we let denote R, resp. U, generic random variables with distributions F_R, resp. F_U.

Let $A(t)$ denote the number of service processes active at time t,

$$A(t) = \int_{\mathcal{R}} \mathbf{1}_{\{s \leq t \leq s+u\}} dN.$$

By definition, $A(t)$ is a Poisson random variable with intensity

$$\mu\left((s, u, r) : s \leq t \leq s + u\right) = \mu\left((s, u, r) : t - u \leq s \leq t\right)$$

$$= \lambda \int_0^\infty \int_0^\infty u F_U(du) F_R(dr)$$

$$= \lambda \int_0^\infty u F_U(du)$$

which is finite iff the expectation of duration time for a service process

$$\mathbb{E}\, U = \int_0^\infty u F_U(du)$$

is finite. If $\mathbb{E}\, U = +\infty$, then $A(t)$ is infinite almost surely.

Recall that a generic Poisson integral

$$\int_R f dN$$

is well defined iff (7.19) holds, i.e. iff

$$\int_\mathcal{R} \min\{|f|, 1\} d\mu < \infty.$$

In particular, for any $\varepsilon > 0$ it is necessary to have $\mu(|f| > \varepsilon) < \infty$. Let us look from this prospective at the instant workload integral (12.1). The necessary assumption for correctness will be

$$\mu\left((s, u, r) : r1_{\{s \le t \le s+u\}} > \varepsilon\right)$$

$$= \lambda \int_0^\infty 1_{\{r > \varepsilon\}} F_R(dr) \cdot \int_0^\infty \left(\int_\mathbb{R} 1_{\{t-u \le s \le t\}} ds\right) F_U(du)$$

$$= \lambda \mathbb{P}(R > \varepsilon) \cdot \int_0^\infty u F_U(du) = \lambda \mathbb{P}(R > \varepsilon) \mathbb{E} U < +\infty.$$

Therefore, excluding the pathological case $\mathbb{P}(R = 0) = 1$, the instant workload makes sense only if

$$\mathbb{E} U = \int_0^\infty u F_U(du) < +\infty.$$

We will also assume that

$$\mathbb{E} R = \int_0^\infty r F_R(dr) < \infty.$$

This is not such a necessary assumption, but if it holds, then we have finite expectations for the instant and for the integral workload, because

$$\mathbb{E} W^\circ(t) = \mathbb{E} \int_\mathcal{R} r1_{\{s \le t \le s+u\}} dN$$

$$= \int_\mathcal{R} r1_{\{s \le t \le s+u\}} d\mu$$

$$= \lambda \int_0^\infty r F_R(dr) \cdot \int_0^\infty \left(\int_{-\infty}^\infty 1_{\{t-u \le s \le t\}} ds\right) F_U(du)$$

$$= \lambda \mathbb{E} R \cdot \int_0^\infty u F_U(du)$$

$$= \lambda \mathbb{E} R \cdot \mathbb{E} U$$

and

$$\mathbb{E} W^*(t) = \int_0^t \mathbb{E} W^\circ(\tau) d\tau = \lambda \mathbb{E} R \cdot \mathbb{E} U \cdot t. \tag{12.5}$$

Before describing the rigorous assumptions on the system recall the standard notation used throughout the chapter. For two functions, f and g we write $f \sim g$ iff

$$\lim_{x \to \infty} \frac{f(x)}{g(x)} = 1.$$

We write $f = o(g)$ iff

$$\lim_{x \to \infty} \frac{f(x)}{g(x)} = 0.$$

Actually we will assume something stronger than just the finite expectations. We suppose that either the variables R and U have finite variance, or their distributions have regular tails. More precisely, either

$$\mathbb{P}(U > u) \sim \frac{c_U}{u^\gamma}, \qquad u \to \infty, \qquad 1 < \gamma < 2, \ c_U > 0, \tag{12.6}$$

or

$$\mathbb{E}U^2 < \infty.$$

In the latter case we formally set $\gamma := 2$.

Notice that if the distribution P_U has a density $f_U(\cdot)$ with asymptotics $f_U(x) \sim \frac{K}{x^{1+\gamma}}$, with $\gamma \in (1, 2)$, then (12.6) holds with $c_U = \frac{k}{\gamma}$.

Analogously, we assume either

$$\mathbb{P}(R > r) \sim \frac{c_R}{r^\delta}, \qquad r \to \infty, \qquad 1 < \delta < 2, \ c_R > 0, \tag{12.7}$$

or

$$\mathbb{E}R^2 < \infty.$$

In the latter case we formally set $\delta := 2$.

Again, if the distribution P_R has a density $f_R(\cdot)$ such that $f_R(x) \sim \frac{K}{x^{1+\delta}}$, with $\delta \in (1, 2)$, then (12.6) holds with $c_R = \frac{k}{\delta}$.

To summarize, the behavior of the service system crucially depends of the parameters $\gamma, \delta \in (1, 2]$.

12.2 *Analysis of workload variance*

Let us start with the instant workload. By using the general formula (7.25) we have

$$\begin{aligned}
\mathbb{V}\mathrm{ar}W^\circ(t) &= \mathbb{V}\mathrm{ar} \int_{\mathcal{R}} r \mathbf{1}_{\{s \leq t \leq s+u\}} dN = \int_{\mathcal{R}} r^2 \mathbf{1}_{\{s \leq t \leq s+u\}} d\mu \\
&= \lambda \int_0^\infty r^2 F_R(dr) \cdot \int_0^\infty \left(\int_{-\infty}^\infty \mathbf{1}_{\{t-u \leq s \leq t\}} ds \right) F_U(du) \\
&= \lambda \, \mathbb{E}R^2 \cdot \int_0^\infty u F_U(du) \\
&= \lambda \, \mathbb{E}R^2 \cdot \mathbb{E}U.
\end{aligned}$$

We observe that the variance is finite whenever R has finite second moment and U has finite expectation.

Similarly, for the covariance of two values of the instant workload taken at different time moments $t_1 \leq t_2$ one obtains by using (7.26)

$$\text{cov}(W^\circ(t_1), W^\circ(t_2)) = \int_{\mathcal{R}} r^2 \mathbf{1}_{\{s \leq t_1 \leq s+u\}} \mathbf{1}_{\{s \leq t_2 \leq s+u\}} d\mu$$

$$= \lambda \int_0^\infty r^2 F_R(dr) \int_0^\infty \left(\int_{-\infty}^\infty \mathbf{1}_{\{t_2 - u \leq s \leq t_1\}} ds \right) F_U(du)$$

$$= \lambda \ \mathbb{E}R^2 \int_{t_2 - t_1}^\infty (u - (t_2 - t_1)) F_U(du). \tag{12.8}$$

Notice that the covariance is always non-negative. The reason for this effect is intuitively clear: there may exist service processes that are active at both times t_1 and t_2, in other words, the processes that start before t_1 and end after t_2. Such processes contribute positively to both instant workloads that we consider.

In a similar way, we may compute and analyze the variance of the integral workload. Namely, by using again the general formula for variance (7.25), we obtain

$$\mathbb{V}ar W^*(t) = \lambda \int_0^\infty r^2 F_R(dr) \int_0^\infty \left(\int_{-\infty}^\infty \ell_t(s, u)^2 ds \right) F_U(du).$$

Fix a variable u and consider the interior expression $\ell_t(s, u)$ as a function of variable s. Two cases appear:

1) $0 \leq u \leq t$. Then $\ell_t(\cdot, u)$ vanishes when $s < -u$ or $s > t$, while ℓ_t varies linearly with unit speed on the intervals $[-u, 0]$ and $[t-u, t]$, and has a constant value u on the interval $[0, t-u]$, see Figure 12.1.

It follows that in this case

$$\int_{-\infty}^\infty \ell_t(s, u)^2 ds = 2 \int_0^u s^2 ds + (t-u)u^2 = tu^2 - \frac{u^3}{3}. \tag{12.9}$$

2) $t \leq u < \infty$. Then $\ell_t(\cdot, u)$ vanishes if $s < -u$ or $s > t$, varies linearly with unit speed on the intervals $[-u, t-u]$ and $[0, t]$, and has a constant value t on the interval $[t-u, 0]$, see Figure 12.2.

It follows that

$$\int_{-\infty}^\infty \ell_t(s, u)^2 ds = 2 \int_0^t s^2 ds + (u-t)t^2 = t^2 u - \frac{t^3}{3}. \tag{12.10}$$

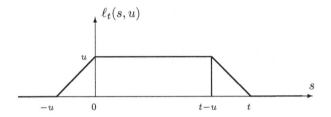

Fig. 12.1 $\ell_t(\cdot, u)$ for the case $u < t$.

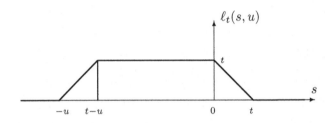

Fig. 12.2 $\ell_t(\cdot, u)$ for the case $t < u$.

We conclude that

$$\mathbb{V}\mathrm{ar}W^*(t) = \lambda \cdot \mathbb{E}R^2 \cdot J, \qquad (12.11)$$

where

$$J := \int_0^t \left(tu^2 - \frac{u^3}{3} \right) F_U(du) + \int_t^\infty \left(t^2 u - \frac{t^3}{3} \right) F_U(du). \qquad (12.12)$$

Again we see that the variance is finite whenever R has finite second moment and U has finite first moment.

We will investigate the behavior of the variance (12.11) when $t \to \infty$. Here, two alternative cases will show up for the first time: the *weak dependence* and the *long range dependence*.

12.2.1   *Weak dependence*

The term *"weak dependence"* vaguely denotes the situation where, while studying some system in a long run, we may neglect the dependence between its instant characteristics at distant moments of time. In particular, the integral characteristics (e.g. the integral workload) of weakly dependent systems essentially behave like sums of i.i.d. random variables, thus their variances will be growing as t when t goes to infinity. Formally, in our case, the *weak dependence* is described by the finiteness of the second moments, i.e. for $\delta = \gamma = 2$, or

$$\mathbb{E}R^2 = \int_0^\infty r^2 F_R(dr) < \infty, \quad \mathbb{E}U^2 := \int_0^\infty u^2 F_U(du) < \infty. \quad (12.13)$$

From four integrals present in (12.12), only one matters for the asymptotic behavior:

$$t \int_0^t u^2 F_U(du) \sim \mathbb{E}U^2 \, t.$$

The other terms are of the order $o(t)$. Indeed,

$$t^3 \int_t^\infty F_U(du) \le t \int_t^\infty u^2 F_U(du) = t \, o(1),$$

$$t^2 \int_t^\infty u F_U(du) \le t \int_t^\infty u^2 F_U(du) = t \, o(1),$$

and for any $\varepsilon > 0$ it is true that

$$\int_0^t u^3 F_U(du) = \int_0^{\varepsilon t} u^3 F_U(du) + \int_{\varepsilon t}^t u^3 F_U(du)$$

$$\le \varepsilon t \, \mathbb{E}U^2 + t \int_{\varepsilon t}^\infty u^2 F_U(du) = \varepsilon t \, \mathbb{E}U^2 + t \, o(1).$$

Therefore, under weak dependence conditions (12.13), we obtain

$$\mathrm{Var} W^*(t) \sim \lambda \, \mathbb{E}R^2 \, \mathbb{E}U^2 \, t, \quad t \to \infty. \quad (12.14)$$

12.2.2 Long range dependence

Formally, the *long range dependence* is described by the parametric assumptions $\delta = 2, 1 < \gamma < 2$, i.e.

$$\mathbb{E}R^2 < \infty, \quad P(U > u) \sim \frac{c_U}{u^\gamma}.$$

In this case, all four integrals that are present in (12.12) have the same order of magnitude. Indeed let

$$\widehat{F}_U(u) := \mathbb{P}(U \ge u)$$

denote the tail of duration distribution. Then, integration by parts and subsequent application of asymptotics (12.6) yield

$$t \int_0^t u^2 F_U(du) = 2t \int_0^t u \widehat{F}_U(u) du - t^3 \widehat{F}_U(t)$$

$$\sim 2c_U \, t \int_0^t u^{1-\gamma} du - c_U \, t^{3-\gamma}$$

$$= c_U \, t^{3-\gamma} \left(\frac{2}{2-\gamma} - 1 \right) = c_U \, t^{3-\gamma} \frac{\gamma}{2-\gamma} ; \quad (12.15)$$

$$t^3 \int_t^\infty F_U(du) = t^3 \widehat{F}_U(t) \sim c_U\, t^{3-\gamma}\ ;$$

$$t^2 \int_t^\infty u F_U(du) = t^2 \int_t^\infty \widehat{F}_U(u)du + t^3 \widehat{F}_U(t)$$

$$\sim c_U\, t^2 \int_t^\infty u^{-\gamma} du + c_U\, t^{3-\gamma}$$

$$= c_U\, t^{3-\gamma} \left(\frac{1}{\gamma-1} + 1 \right) = c_U\, t^{3-\gamma}\, \frac{\gamma}{\gamma-1}\ ;\quad (12.16)$$

and

$$\int_0^t u^3 F_U(du) = 3 \int_0^t u^2 \widehat{F}_U(u)du - t^3 \widehat{F}_U(t)$$

$$\sim 3c_U \int_0^t u^{2-\gamma} du - c_U\, t^{3-\gamma}$$

$$= c_U\, t^{3-\gamma} \left(\frac{3}{3-\gamma} - 1 \right) = c_U\, t^{3-\gamma}\, \frac{\gamma}{3-\gamma}\ .$$

We plug in (12.12) the constants that appeared above and obtain:

$$\frac{\gamma}{2-\gamma} - \frac{\gamma}{3(3-\gamma)} + \frac{\gamma}{\gamma-1} - \frac{1}{3} = \frac{\gamma}{(2-\gamma)(\gamma-1)} - \frac{1}{3-\gamma}$$

$$= \frac{2}{(2-\gamma)(\gamma-1)(3-\gamma)}\ .$$

Therefore, under long range dependence it is true that

$$\mathbb{V}ar W^*(t) \sim \lambda\, \mathbb{E}R^2\, \frac{2c_U}{(2-\gamma)(\gamma-1)(3-\gamma)}\, t^{3-\gamma}, \qquad t \to \infty. \quad (12.17)$$

The same calculations provide a useful *uniform* bound

$$\mathbb{V}ar W^*(t) \leq \widehat{c_U}\lambda\, \mathbb{E}R^2\ t^{3-\gamma} \qquad\qquad (12.18)$$

valid for some constant $\widehat{c_U} > 0$ and all $t > 0$.

Not surprisingly, as γ goes to 2, the exponent $3-\gamma$ goes to 1, approaching weak dependence case with its linear growth of the variance. Yet for any $\gamma < 2$ the variance grows faster than linearly due to the important positive dependence between the values of instant workload.

Recall that we have observed the same power growth of variance for H-fractional Brownian motion, $\mathbb{V}ar B^H(t) = |t|^{2H}$, see (6.12). This is a first hint indicating that in a long run the integral workload of a system exhibiting long range dependence may behave like H-fractional Brownian motion with $H = \frac{3-\gamma}{2}$. Notice that $H \in (\frac{1}{2}, 1)$ whenever $\gamma \in (1, 2)$. This is not surprising because we know that the increments of the workload are positively correlated, which is also true for fractional Brownian motion exactly when $H \in (\frac{1}{2}, 1]$. We refer to [3] for one of few models where H-fractional Brownian motion with $H \in (0, \frac{1}{2})$ shows up in the limit.

13 Limit Theorems for the Workload

13.1 *Centered and scaled workload process*

In this section, we will explore the behavior of the integral workload as a process (function of time) observed on the long time intervals. Exactly as in the case of limit theorems for sums (see Section 3), in the long run particular features disappear and some kind of invariant behavior emerges.

In order to obtain a meaningful limit, one must do the following preliminary operations.

- scale (contract) the time so that it would run through the standard time interval;
- center the workload process;
- divide it by an appropriate scalar factor.

We choose $[0, 1]$ as a standard time interval. Since we will study the workload on the long time interval $[0, a]$ with $a \to \infty$, the time scaling can be written as $W^*(at), t \in [0, 1]$, so that when t runs through $[0, 1]$ the workload argument runs through $[0, a]$.

Centering and scaling by appropriate factor b lead to a *normalized integral workload process*

$$Z_a(t) := \frac{W^*(at) - \mathbb{E}R \cdot \mathbb{E}U \cdot a\lambda t}{b} , \qquad t \in [0, 1]. \qquad (13.1)$$

Recall that

$$\mathbb{E} W^*(at) = \mathbb{E}R \cdot \mathbb{E}U \cdot a\lambda t$$

by (12.5), thus $Z_a(t)$ is centered.

In the following we will consider the observation horizon length a and arrival intensity λ as the variables (at least one of them must tend to infinity, in order to provide us with many observed service processes), while the space scaling factor b depends on these variables and the form of this dependence is determined by the system parameters γ, δ. Thus b is just an abbreviation for $b(a, \lambda)$.

For example, in the simplest case of rather short and not very intensive service processes ($\gamma = \delta = 2$) we have $b \approx (a\lambda)^{1/2}$, square root of the variance (12.14), in complete analogy to the Lévy's central limit theorem 3.1 for the i.i.d. variables with finite variance.

Our aim is to find a process to which the normalized workload Z_a converges when a tends to infinity. The limiting process may vary from one situation to another but it certainly has *stationary increments* because the integral workload $W^*(\cdot)$ has this property, and linear transformation leading from $W^*(\cdot)$ to $Z_a(\cdot)$ does not destroy it.

Therefore, Wiener process, other fractional Brownian motions, Lévy processes are among reasonable candidates for being a limit for Z_a.

It is much less obvious whether the limiting process should have *independent increments*. As we know from (12.8), there is a positive covariance between the values of the instant workload. For its integral W^* it means the positive dependence of increments, which is inherited by $Z_a(\cdot)$. This dependence may vanish or not when we pass to the limit. In the first case we would obtain a Wiener process or some other Lévy process, while in the second case we could obtain, for example, an H-fractional Brownian motion with $H > \frac{1}{2}$. Recall that $H < \frac{1}{2}$ corresponds to negatively dependent increments, which is not compatible with the workload's properties. Interesting examples of the limiting processes with dependent increments other than fractional Brownian motion also exist, see Subsection 13.5 below.

Whether the increments of the limiting process are dependent or not, is determined very much by the role played by the "long" service processes, i.e. those having duration u comparable with observation horizon a. If their contribution to the integral workload $W^*(a)$ is asymptotically negligible, we have independent increments in the limit, otherwise the long processes render dependent the increments of the limiting process.

The continuity of the limiting process is another interesting issue for discussion. The jumps, if they exist, appear when there are "short" service processes ("short" means that the duration of service u is much smaller than observation horizon a) whose contribution $r\ell_{at}(s, u)$ to the workload $W^*(a)$ is comparable to the size of this workload. Sometimes, "short" processes are not so important and we obtain a continuous process in the limit, e.g. a Wiener process (Subsection 13.2) or a fractional Brownian motion (Subsection 13.3). Sometimes the jumps do appear, e.g. when the limiting process is a stable Lévy process, or one of the Telecom processes, see Subsections 13.4 and 13.5 below.

Another typical property of a limiting process is self-similarity. Assume that the norming factor $b = b(a, \lambda)$ is regularly varying, i.e. for any $k > 0$

$$\frac{b(k\,a, \lambda)}{b(a, \lambda)} \to k^{\alpha}. \tag{13.2}$$

In all numerous subsequent cases this property is verified.

For any $c, a > 0$ we have an identity

$$Z_{a/c}(ct) = \frac{W^*(at) - \mathbb{E}\,W^*(at)}{b(a/c, \lambda)}$$

$$= \frac{b(a, \lambda)}{b(a/c, \lambda)} \, \frac{W^*(at) - \mathbb{E}\,W^*(at)}{b(a, \lambda)}$$

$$= \frac{b(a, \lambda)}{b(a/c, \lambda)} \, Z_a(t). \tag{13.3}$$

If a convergence of normalized workload holds, $Z_a \overset{\text{f.d.d.}}{\longrightarrow} Y$, then we may pass to the limit in (13.3), using (13.2) with $k = c^{-1}$ and obtain the self-similarity property

$$P_{Y(ct)} = P_{c^\alpha Y(t)}$$

for any fixed t, as well for all finite-dimensional distributions of the process Y.

Now we present a panorama of concrete results for specific situations. The general map is given by Figures 13.1–13.2. There are several zones of parameters, each zone corresponding to a particular limit process. We handle most of them in this section. Note that one observes a similar picture in non-Poissonian models, e.g. such as on–off models [27].

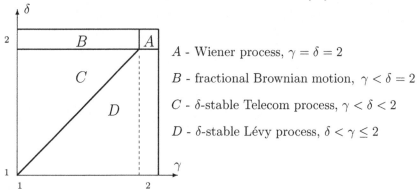

A - Wiener process, $\gamma = \delta = 2$

B - fractional Brownian motion, $\gamma < \delta = 2$

C - δ-stable Telecom process, $\gamma < \delta < 2$

D - δ-stable Lévy process, $\delta < \gamma \leq 2$

Fig. 13.1   Limit theorems for high intensity.

13.2   Weak dependence: convergence to Wiener process

We will search a limit theorem for Z_a in the case $\gamma = \delta = 2$ and choose the scaling factor such that $Z_a(1)$ would have asymptotically unit variance,

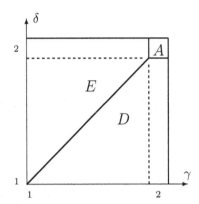

A - Wiener process, $\gamma = \delta = 2$

E - γ-stable Lévy process, $\gamma < \delta \leq 2$

D - δ-stable Lévy process, $\delta < \gamma \leq 2$

Fig. 13.2   Limit theorems for low intensity.

i.e., according to (12.14),

$$1 \sim \frac{\mathbb{V}arW^*(a)}{b^2} \sim \frac{\mathbb{E}R^2 \ \mathbb{E}U^2 \ a\lambda}{b^2}.$$

Therefore, we may let

$$b := \left(\mathbb{E}R^2 \ \mathbb{E}U^2 \ a\lambda\right)^{1/2}. \tag{13.4}$$

Such scaling yields by (12.14), for every $t \in [0,1]$

$$\mathbb{V}arZ_a(t) = \frac{\mathbb{V}arW^*(at)}{b^2} \sim \frac{\mathbb{E}R^2 \ \mathbb{E}U^2 \ a\lambda t}{b^2} = t, \qquad \text{as } a \to \infty. \tag{13.5}$$

Also recall that for any $t_1, t_2 \in [0,1]$ with $t_1 \leq t_2$ it is true that

$$\mathbb{V}ar(Z_a(t_2) - Z_a(t_1)) = b^{-2} \ \mathbb{V}ar \left(W^*(t_2) - W^*(t_1)\right)$$

$$= b^{-2} \ \mathbb{V}ar \int_{t_1}^{t_2} W^\circ(\tau)d\tau$$

$$= b^{-2} \ \mathbb{V}ar \int_0^{t_2-t_1} W^\circ(\tau)d\tau$$

$$= b^{-2} \ \mathbb{V}arW^*(t_2 - t_1)$$

$$= \mathbb{V}arZ_a(t_2 - t_1) \tag{13.6}$$

$$\to t_2 - t_1.$$

Furthermore, by using the identity

$$\mathbb{V}ar(Z_a(t_2) - Z_a(t_1)) = \mathbb{V}arZ_a(t_1) + \mathbb{V}arZ_a(t_2) - 2\operatorname{cov}(Z_a(t_1), Z_a(t_2))$$

we derive

$$\operatorname{cov}(Z_a(t_1), Z_a(t_2))$$

$$= \frac{1}{2} \left(\mathbb{V}arZ_a(t_1) + \mathbb{V}arZ_a(t_2) - \mathbb{V}ar(Z_a(t_2) - Z_a(t_1))\right) \tag{13.7}$$

$$\to \frac{1}{2} \left(t_1 + t_2 - (t_2 - t_1)\right) = t_1.$$

Thus,

$$\text{cov}(Z_a(t_1), Z_a(t_2)) \to \min(t_1, t_2).$$

Therefore, the limiting covariance is that of a Wiener process.

Normalized workload process Z_a similar to a Wiener process is represented on Figure 13.3.

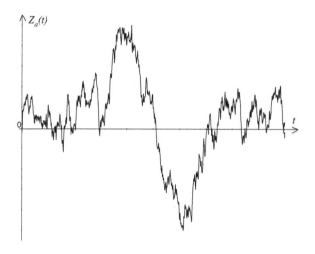

Fig. 13.3 Normalized workload Z_a similar to a Wiener process.

We will now confirm the guess by proving that the Wiener process indeed is a required limit.

Theorem 13.1. *Let $\gamma = \delta = 2$ and assume that $a \to \infty$ and $a\lambda \to \infty$. Define the scaling factor by (13.4). Let W be a Wiener process. Then we have*

$$Z_a \Rightarrow W \ \text{in} \ \mathbb{C}[0,1].$$

In particular,

$$Z_a \overset{f.d.d.}{\longrightarrow} W.$$

Proof.

Step 1: convergence of one-dimensional distributions.

In order to prove that for each $t \geq 0$

$$Z_a(t) \Rightarrow W(t),$$

or, in other words, $P_{Z_a(t)} \Rightarrow \mathcal{N}(0, t)$, we will use Proposition 8.1. We already know from (13.5) that $\mathbb{V}ar\,Z_a(t) \to t$, as $a \to \infty$; therefore, it only remains to verify the second assumption (8.3) of that proposition. Recall that by (12.3)

$$Z_a(t) = \int_{\mathcal{R}} \frac{r\ell_{at}(s, u)}{(a\lambda \mathbb{E}R^2\,\mathbb{E}U^2\,)^{1/2}}\,d\tilde{N},$$

where the kernel $\ell_t(s, u)$ was defined in (12.4) and $\tilde{N}$ is a centered version of Poisson measure N. Recall also that notation $\gamma = \delta = 2$ just means $\mathbb{E}R^2 < \infty$, $\mathbb{E}U^2 < \infty$, which makes meaningful the expression given above.

We must check that for any $\varepsilon > 0$ it is true that

$$\int_{\{r\ell_{at}/(a\lambda)^{1/2}>\varepsilon\}} \frac{r^2\ell_{at}(s, u)^2}{a\lambda} F_R(dr)F_U(du)\lambda ds \to 0.$$

Let us first get rid of the zone $\{(s, u, r) : u > at\}$. In fact, we have already seen that its contribution to the variance is negligible since by (12.9)

$$\int_{-\infty}^{\infty} \ell_{at}(s, u)^2 ds = (at)^2 u - \frac{(at)^3}{3} \leq (at)^2 u\ ,$$

hence,

$$\int_{\{u>at\}} \frac{r^2\ell_{at}(s, u)^2}{a\lambda} F_R(dr)F_U(du)\lambda ds \leq \mathbb{E}R^2\,a^{-1} \int_{at}^{\infty} (at)^2 u F_U(du)$$

$$\leq \mathbb{E}R^2\,t \int_{at}^{\infty} u^2 F_U(du) \to 0.$$

Next, let us consider the zone $\{(s, u, r) : u \leq at\}$, where by (12.10)

$$\int_{-\infty}^{\infty} \ell_{at}(s, u)^2 ds = (at)u^2 - u^3/3 \leq (at)u^2. \tag{13.8}$$

Moreover, using a trivial inequality $\ell_{at}(s, u) \leq u$, we can cover the zone $\{(s, u, r) : r\ell_{at}(s, u)/(a\lambda)^{1/2} > \varepsilon\}$ we are interested in by a larger zone $\{(s, u, r) : ru/(a\lambda)^{1/2} > \varepsilon\}$, and cover the latter by the union of two simpler zones $\{(s, u, r) : r > (a\lambda\varepsilon^2)^{1/4}\}$ and $\{(s, u, r) : u > (a\lambda\varepsilon^2)^{1/4}\}$. In this way, we obtain the estimate

$$\int_{\left\{\substack{r\ell_{at}/(a\lambda)^{1/2}>\varepsilon \\ u\leq at}\right\}} \frac{r^2\ell_{at}(s, u)^2}{a\lambda}\,d\mu$$

$$\leq \int_{\left\{\substack{r>(a\lambda\varepsilon^2)^{1/4} \\ u\leq at}\right\}} \frac{r^2\ell_{at}(s, u)^2}{a\lambda}\,d\mu + \int_{\left\{\substack{u>(a\lambda\varepsilon^2)^{1/4} \\ u\leq at}\right\}} \frac{r^2\ell_{at}(s, u)^2}{a\lambda}\,d\mu.$$

For the first integral by using (13.8) we obtain the bound

$$\int_{(a\lambda\varepsilon^2)^{1/4}}^{\infty} r^2 F_R(dr) \cdot (a\lambda)^{-1} \lambda \int_0^{at} \left(\int_{-\infty}^{\infty} \ell_{at}(s,u)^2 ds \right) F_U(du)$$
$$\leq \int_{(a\lambda\varepsilon^2)^{1/4}}^{\infty} r^2 F_R(dr) \cdot a^{-1}(at) \int_{\mathbb{R}} u^2 F_U(du) \to 0,$$

due to the first factor. We used here theorem's assumptions: $\mathbb{E}R^2 < \infty$, $\mathbb{E}U^2 < \infty$, and $a\lambda \to \infty$.

For the second integral we obtain, still by using (13.8), the bound

$$\int_{\mathbb{R}} r^2 F_R(dr) \cdot (a\lambda)^{-1} \lambda \int_{(a\lambda\varepsilon^2)^{1/4}}^{at} \left(\int_{-\infty}^{\infty} \ell_{at}(s,u)^2 ds \right) F_U(du)$$
$$\leq \int_{\mathbb{R}} r^2 F_R(dr) \cdot a^{-1}(at) \int_{(a\lambda\varepsilon^2)^{1/4}}^{at} u^2 F_U(du)$$
$$\leq \int_{\mathbb{R}} r^2 F_R(dr) \cdot t \cdot \int_{(a\lambda\varepsilon^2)^{1/4}}^{\infty} u^2 F_U(du) \to 0,$$

due to the last factor. Again, we used here the theorem's assumptions $\mathbb{E}R^2 < \infty$, $\mathbb{E}U^2 < \infty$, and $a\lambda \to \infty$.

Step 2: convergence of distributions for increments.

Recall that Z_a inherits from W^* the property to have stationary increments. This means that for any non-negative $t_1 \leq t_2$, the increment $Z_a(t_2) - Z_a(t_1)$ is equidistributed with $Z_a(t_2 - t_1)$. Therefore, we may restate the result of Step 1 as

$$Z_a(t_2) - Z_a(t_1) \Rightarrow W(t_2) - W(t_1), \tag{13.9}$$

or $P_{Z_a(t_2)-Z_a(t_1)} \Rightarrow \mathcal{N}(0, t_2 - t_1)$.

Step 3: convergence of multivariate distributions.

We have to show that for any $n \geq 2$ and any $0 \leq t_1 < \cdots < t_n$ the weak convergence of finite-dimensional distributions holds

$$P_{t_1,\ldots,t_n}^{Z_a} \Rightarrow P_{t_1,\ldots,t_n}^W, \qquad \text{as } a\lambda \to \infty.$$

This is equivalent (cf. Exercise 4.7) to proving the convergence for the vectors of increments, i.e.

$$\Delta_a \Rightarrow \Delta_W, \tag{13.10}$$

where

$$\Delta_a := (Z_a(t_j) - Z_a(t_{j-1}))_{j=1}^n, \qquad \Delta_W := (W(t_j) - W(t_{j-1}))_{j=1}^n, \tag{13.11}$$

and we let $t_0 := 0$.

Notice that by (13.9) we have component-wise convergence

$$(\Delta_a)_j \Rightarrow (\Delta_W)_j = W(t_j) - W(t_{j-1}), \qquad j = 1, 2, \ldots \qquad (13.12)$$

Would the components $(\Delta_a)_j$ be independent, this would yield the required convergence (13.10) by virtue of Exercise 4.6. But they *are* dependent! Therefore, we must go around the obstacle by splitting Δ_a into two parts,

$$\Delta_a =: \Delta'_a + \Delta''_a, \qquad (13.13)$$

so that: a) the components $(\Delta'_a)_j$ of the main part are independent, and b) for the remainder we have $\Delta''_a \Rightarrow 0$.

Then, since the difference $(\Delta''_a)_j$ between $(\Delta_a)_j$ and $(\Delta'_a)_j$ is small, convergence (13.12) implies (cf. Exercise 1.19) that $(\Delta'_a)_j \Rightarrow (\Delta_W)_j$. This time component-wise convergence criterion from Exercise 4.6 applies correctly and we get

$$\Delta'_a \Rightarrow \Delta_W. \qquad (13.14)$$

Since the difference Δ''_a between Δ_a and Δ'_a is small, convergence (13.10) follows (cf. Exercise 4.4) from (13.14).

To achieve the desired splitting (13.13), let

$$(\Delta_a)_j = Z_a(t_j) - Z_a(t_{j-1}) = \int_{\mathcal{R}} \frac{r(\ell_{at_j}(s,u) - \ell_{at_{j-1}}(s,u))}{b} \, d\widetilde{N}$$

$$:= \int_{\mathcal{R}} \frac{r\widetilde{\ell}_j(a,s,u)}{b} \, d\widetilde{N}$$

$$= \int_{\{s \in (at_{j-1}, at_j]\}} \frac{r\widetilde{\ell}_j(a,s,u)}{b} \, d\widetilde{N} + \int_{\{s \leq at_{j-1}\}} \frac{r\widetilde{\ell}_j(a,s,u)}{b} \, d\widetilde{N}$$

$$:= (\Delta'_a)_j + (\Delta''_a)_j. \qquad (13.15)$$

Here

$$\widetilde{\ell}_j(a,s,u) = \ell_{at_j}(s,u) - \ell_{at_{j-1}}(s,u) = \Big| [s, s+u] \cap (at_{j-1}, at_j] \Big|.$$

The components of vector Δ'_a are centered Poisson integrals of disjointly supported functions. Hence they are independent, as required. When evaluating the components $(\Delta''_a)_j$, by stationarity we may only consider $j = 1$, where we have

$$\mathbb{E}\,(\Delta_a'')_1^2 = \int_{\{s \le 0\}} \frac{r^2 \ell_{at_1}(s, u)^2}{b^2}\, d\mu$$

$$\le \int_{\mathcal{R}} \frac{r^2 \min(u, at_1)^2}{b^2} \mathbf{1}_{\{-u \le s \le 0\}}\, d\mu$$

$$\le \frac{\lambda\,\mathbb{E} R^2}{b^2} \int_0^{\infty} \left(\int_{-u}^0 \min(u, a)^2 ds \right) F_U(du)$$

$$= \frac{1}{\mathbb{E} U^2\, a} \int_0^{\infty} u \min(u, a)^2 F_U(du).$$

Now take any $\varepsilon > 0$ and use the estimate

$$\int_0^{\infty} u \min(u, a)^2 F_U(du) \le \varepsilon a \int_0^{\varepsilon a} u^2 F_U(du) + a \int_{\varepsilon a}^{\infty} u^2 F_U(du).$$

It follows that

$$\mathbb{E}\,(\Delta_a'')_1^2 \le \frac{1}{\mathbb{E} U^2} \left(\varepsilon \mathbb{E} U^2 + \int_{\varepsilon a}^{\infty} u^2 F_U(du) \right).$$

Since ε can be chosen arbitrarily small, we obtain

$$\lim_a \mathbb{E}\,(\Delta_a'')_1^2 = 0. \tag{13.16}$$

Hence the random vectors Δ_a'' go to zero in probability and we have proved all required properties of the splitting (13.13).

Remark 13.2. Notice that (13.16) is a proof's key point where we show that in the setting of Theorem 13.1 "long" service processes are not enough substantial to provide the dependence of workload on the distant intervals of time.

Step 4: weak convergence in $\mathbb{C}[0, 1]$. Assume first[3] that U and R are bounded: with probability one $U \le U_0$ and $R \le R_0$. Then we have the following bound for $\mathbb{E}\, Z_a(t)^4$. By using (2.7),

$$\mathbb{E}\, Z_a(t)^4 = \int_{\mathcal{R}} \left(\frac{r \ell_{at}(s, u)}{b} \right)^4 d\mu + 3 \left(\int_{\mathcal{R}} \left(\frac{r \ell_{at}(s, u)}{b} \right)^2 d\mu \right)^2$$

$$\le c_1 \int_0^{U_0} \frac{\min(u^4, (at)^4)(u + at)}{a^2 \lambda} F_U(du)$$

$$+ c_2 \left(\int_0^{U_0} \frac{\min(u^2, (at)^2)(u + at)}{a} F_U(du) \right)^2$$

---

[3]This part of the proof operates with more involved arguments and may be omitted at the first reading.

where $c_1 := \frac{R_0^4}{\mathbb{E}R^2\,\mathbb{E}U^2}$, $c_2 = \frac{3}{(\mathbb{E}U^2)^{1/2}}$. Assuming that $at \geq U_0$, we obtain

$$\mathbb{E}\,Z_a(t)^4 \leq 2c_1 U_0^4\frac{t}{a\lambda} + c_2\left(2U_0^2 t\right)^2 := c_3\frac{t}{a\lambda} + c_4 t^2.$$

Next, we fix ε, η and try to find a small $t = t(\varepsilon, \eta)$ such that for all a large enough

$$\mathbb{P}\left(\sup_{0 \leq \tau \leq t} |Z_a(\tau)| \geq \varepsilon\right) \leq \eta\,t. \tag{13.17}$$

Indeed, the process $Z_a(\cdot)$ is associated (cf. [15], Section 1.3) and we may apply a continuous time version of the Newman-Wright maximal inequality for sums of associated variables (see [73, 74] or [15], Theorem 4.6)

$$\mathbb{P}\left(\sup_{0 \leq \tau \leq t} |Z_a(\tau)| \geq q(\mathbb{E}\,Z_a(t)^2)^{1/2}\right) \leq 2\,\mathbb{P}\left(|Z_t| \geq (q - \sqrt{2})(\mathbb{E}\,Z_a(t)^2)^{1/2}\right),$$

which is valid for any $q > 0$. We may rewrite this as

$$\mathbb{P}\left(\sup_{0 \leq \tau \leq t} |Z_a(\tau)| \geq \varepsilon\right) \leq 2\,\mathbb{P}\left(|Z_a(t)| \geq \frac{q - \sqrt{2}}{q}\,\varepsilon\right), \tag{13.18}$$

where

$$q := \frac{\varepsilon}{(\mathbb{E}\,Z_a(t)^2)^{1/2}}.$$

Recall that $\lim_a \mathbb{E}\,Z_a(t)^2 = t$. Therefore, if t is so small that $\varepsilon > 2\sqrt{t}$, for large a we have $q > 2$, hence $\frac{q-\sqrt{2}}{q} > \frac{1}{4}$ and

$$\mathbb{P}\left(\sup_{0 \leq \tau \leq t} |Z_a(\tau)| \geq \varepsilon\right) \leq 2\,\mathbb{P}(|Z_a(t)| \geq \varepsilon/4) \leq 2^9\,\varepsilon^{-4}\,\mathbb{E}\,Z_a(t)^4$$

$$\leq 2^9\,\varepsilon^{-4}\left(c_3\frac{t}{a\lambda} + c_4 t^2\right).$$

Since $a\lambda \to \infty$, we may choose t so small that for large a one has

$$2^9\,\varepsilon^{-4}\left(\frac{c_3}{a\lambda} + c_4 t\right) \leq \eta,$$

and we obtain (13.17). Finally, since the workload has stationary increments, we may rewrite (13.17) in a uniform fashion

$$\sup_{0 \leq \theta \leq 1} \mathbb{P}\left(\sup_{\substack{\theta \leq \tau \leq \theta+t \\ \tau \leq 1}} |Z_a(\tau) - Z_a(\theta)| \geq \varepsilon\right) \leq \eta\,t,$$

which matches the weak convergence criterion (11.6).

For dropping unnecessary assumptions $U \leq U_0$ and $R \leq R_0$, one may first fix some large $U_0, R_0 > 0$ and split the workload process $W(\cdot)$ into two independent processes $W := W_0 + W^\dagger$, where the main part W_0 corresponds to the service processes (s, u, r) such that $u \leq U_0$ and $r \leq R_0$ while the remainder $W^\dagger$ corresponds to the exceptional service processes with either $u > U_0$ or $r > R_0$. The normalized workload also splits into two parts $Z_a := Z_{a,0} + Z_a^\dagger$. We may apply the proved result to $Z_{a,0}$ and obtain convergence to a slightly scaled Wiener process. The remainder $Z_a^\dagger$ turns out to be uniformly small in probability. Indeed, notice that uniformly in a

$$D = D(U_0, R_0) := \mathbb{E}\, Z_a^\dagger(1)^2 \leq \frac{c\, \mathbb{E}\, R^2 U^2 \mathbf{1}_{\{U > U_0 \text{ or } R > R_0\}}}{\mathbb{E} R^2 \, \mathbb{E} U^2} \to 0,$$

as $U_0, R_0 \nearrow +\infty$. Now by (13.18) we have for any $\varepsilon > 2D^{1/2}$

$$\mathbb{P}\left(\sup_{0 \leq \tau \leq 1} |Z_a^\dagger(\tau)| \geq \varepsilon \right) \leq 2\,\mathbb{P}\left(|Z_a^\dagger(1)| \geq \varepsilon/4 \right) \leq 2^5 \varepsilon^{-2} D \to 0,$$

whenever ε is fixed and $U_0, R_0 \nearrow +\infty$. The few remaining formal calculations are left to the reader as an exercise. $\qquad\square$

Exercise 13.3. One of the questionable assumptions of the considered model is independence of U and R. In case of weak dependence this assumption can be dropped. Namely, replace the definition of intensity measure by

$$\mu(ds, du, dr) = \lambda ds F_{UR}(du, dr).$$

Instead of $\mathbb{E} U^2 < \infty$, $\mathbb{E} R^2 < \infty$, assume only that

$$\mathbb{E}\,(U^2 R^2) = \int u^2 r^2 F_{UR}(du, dr) < \infty.$$

In the definition (13.1) of $Z_a(t)$ one should of course replace the centering term by $\mathbb{E}\,(RU) \cdot a\lambda \cdot t$. Prove a complete analogue of Theorem 13.1 with scaling (13.4) replaced with

$$b := \left(\mathbb{E}\,(R^2 U^2)\, a\, \lambda \right)^{1/2}.$$

13.3 Long range dependence: convergence to fBm

Long range dependence appears when we have sufficiently many "long" service processes. Here "long" means "having duration u comparable to the observation horizon a". Let us fix an $h \in (0, 1]$ and consider the number $Q_{a,h}$ of service processes covering both times 0 and ha. This is a Poisson random variable

$$Q_{a,h} := \int_{\mathcal{R}} \mathbf{1}_{\{s \leq 0 \leq ah \leq u + s\}} dN$$

with expectation

$$\mathbb{E}\,Q_{a,h} = \int_{\mathcal{R}} 1_{\{s\leq 0\leq ah\leq u+s\}}d\mu$$

$$= \lambda \int_0^\infty \left(\int_0^\infty 1_{\{u\geq ah+\tilde{s}\}}d\tilde{s}\right) F_U(du)$$

$$= \lambda \int_0^\infty \left(\int_0^\infty 1_{\{u\geq ah+\tilde{s}\}}F_U(du)\right) d\tilde{s}$$

$$= \lambda \int_0^\infty \mathbb{P}(U\geq ah+\tilde{s})d\tilde{s} = \lambda \int_{ah}^\infty \mathbb{P}(U\geq v)dv$$

$$\sim \lambda \int_{ah}^\infty c_U v^{-\gamma}dv = \frac{c_U \lambda}{(\gamma-1)(ah)^{\gamma-1}},$$

whenever $\gamma < 2$, i.e. (12.6) holds. Therefore, if we wish to have *many* "long" processes bringing long range dependence, we must impose the assumption

$$\frac{\lambda}{a^{\gamma-1}} \to \infty \qquad (13.19)$$

called *high intensity condition*. We will see below that the limiting behavior of the workload is substantially different when this condition holds or fails, compare Figures 13.1 and 13.2.

We will search a limit theorem for Z_a in the case $1 < \gamma < 2, \delta = 2$ and choose again the scaling factor b such that $Z_a(1)$ would have asymptotically unit variance. Let denote $H = \frac{3-\gamma}{2}$. According to (12.17),

$$1 \sim \frac{\mathbb{V}arW^*(a)}{b^2} \sim \frac{2c_U}{(2-\gamma)(\gamma-1)(3-\gamma)} \frac{\lambda\,\mathbb{E}R^2\ a^{2H}}{b^2}.$$

Therefore, we may let

$$b := \left(\frac{2c_U\lambda\mathbb{E}R^2}{(2-\gamma)(\gamma-1)(3-\gamma)}\right)^{1/2} a^H. \qquad (13.20)$$

Such scaling yields by (12.17), for every $t \in [0,1]$

$$\mathbb{V}arZ_a(t) = \frac{\mathbb{V}arW^*(at)}{b^2}$$

$$\sim \frac{2c_U\lambda\ \mathbb{E}R^2\ (at)^{3-\gamma}}{(2-\gamma)(\gamma-1)(3-\gamma)b^2} = t^{2H}, \qquad \text{as } a\to\infty. \quad (13.21)$$

Again for all $t_1, t_2 \in [0,1]$ with $t_1 \leq t_2$ we apply (13.6)

$$\mathbb{V}ar(Z_a(t_2) - Z_a(t_1)) = \mathbb{V}ar Z_a(t_2 - t_1) \to (t_2 - t_1)^{2H}.$$

Finally, it follows from the identity (13.7) that

$$\mathrm{cov}(Z_a(t_1), Z_a(t_2)) \to \frac{1}{2}\left(t_1^{2H} + t_2^{2H} - (t_2 - t_1)^{2H}\right). \tag{13.22}$$

Therefore, the limiting covariance is that of fractional Brownian motion.

Normalized workload process Z_a similar to a fractional Brownian motion is represented on Figure 13.4.

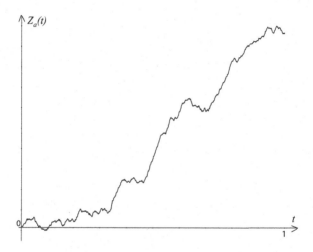

Fig. 13.4   Normalized workload process Z_a similar to a fractional Brownian motion.

We will now confirm the guess by proving that H-fractional Brownian motion indeed is a required limit.

Theorem 13.4. *Let $\gamma \in (1,2)$ and $\delta = 2$. Assume that $a \to \infty$ and that high intensity condition (13.19) holds.*

Define the scaling factor by (13.20). Let B^H be an H-fractional Brownian motion with $H = \frac{3-\gamma}{2}$. Then we have

$$Z_a \Rightarrow B^H \text{ in } \mathbb{C}[0,1].$$

In particular, $Z_a \xrightarrow{f.d.d.} B^H$.

Proof.   *Step 1: convergence of one-dimensional distributions.*

In order to prove that for each $t \geq 0$

$$Z_a(t) \Rightarrow B^H(t),$$

or, in other words, $P_{Z_a(t)} \Rightarrow \mathcal{N}(0, t^{2H})$, we will use Proposition 8.1. We already know from (13.21) that $\mathbb{V}\mathrm{ar}\, Z_a(t) \to t^{2H}$, as $a \to \infty$; therefore, it only remains to verify the second assumption (8.3) of that proposition.

Recall that by (12.3)

$$Z_a(t) = \int_{\mathcal{R}} \frac{r\ell_{at}(s, u)}{(K\lambda)^{1/2} a^{(3-\gamma)/2}} \, d\widetilde{N}, \qquad (13.23)$$

where the kernel $\ell_t(\cdot, \cdot)$ is defined in (12.4), $\widetilde{N}$ is a centered version of Poisson measure N and $K := \frac{2c_U \mathbb{E} R^2}{(2-\gamma)(\gamma-1)(3-\gamma)}$ is a constant which is unimportant at this point. Let

$$f_a(s, u, r) := \frac{r\ell_{at}(s, u)}{a^{(3-\gamma)/2}\lambda^{1/2}} \ .$$

Checking (8.3) means to verify that for any $\varepsilon > 0$ it is true that

$$\int_{\{f_a > \varepsilon\}} f_a^2 F_R(dr) F_U(du) \lambda ds \to 0. \qquad (13.24)$$

Fix a large $M > 0$ and split the integral into two parts:

$$\int_{\{f_a > \varepsilon\}} f_a^2 d\mu = \int_{\{\substack{f_a > \varepsilon \\ r \leq M}\}} f_a^2 d\mu + \int_{\{\substack{f_a > \varepsilon \\ r > M}\}} f_a^2 d\mu \leq \int_{\{\substack{f_a > \varepsilon \\ r \leq M}\}} f_a^2 d\mu + \int_{\{r > M\}} f_a^2 d\mu.$$

We show now that the first integral simply vanishes for large a. Indeed, under assumption (13.19) we have, for the service processes (s, u, r) such that $r \leq M$,

$$f_a(s, u, r) = \frac{r\ell_{at}(s, u)}{a^{(3-\gamma)/2}\lambda^{1/2}} \leq \frac{Mat}{a^{(3-\gamma)/2}\lambda^{1/2}} = Mt \left(\frac{a^{\gamma-1}}{\lambda}\right)^{1/2} \to 0.$$

Now we consider the second integral by using the bounds for the variance from Subsection 12.2. Since $\gamma \in (1, 2)$, we may use first (12.9) and (12.10), then (12.15) and (12.16), which leads to

$$\int_{\{r > M\}} f_a^2 d\mu = \frac{\mathbb{E}\left(R^2 \mathbf{1}_{\{R > M\}}\right)}{a^{3-\gamma}} \left(\int_0^{at} + \int_{at}^{\infty}\right) \left(\int_{\mathbb{R}} \ell_{at}(s, u)^2 ds\right) F_U(du)$$

$$\leq \frac{\mathbb{E}\left(R^2 \mathbf{1}_{\{R > M\}}\right)}{a^{3-\gamma}} \left(at \int_0^{at} u^2 F_U(du) + (at)^2 \int_{at}^{\infty} u F_U(du)\right)$$

$$\sim \frac{\mathbb{E}\left(R^2 \mathbf{1}_{\{R > M\}}\right)}{a^{3-\gamma}} \left(\frac{c_U \gamma (at)^{3-\gamma}}{2 - \gamma} + \frac{c_U \gamma (at)^{3-\gamma}}{\gamma - 1}\right)$$

$$\sim \frac{c_U \gamma t^{3-\gamma} \mathbb{E}\left(R^2 \mathbf{1}_{\{R > M\}}\right)}{(2 - \gamma)(\gamma - 1)}.$$

By choosing M to be large enough, we may render this expression arbitrarily small, thus (13.24) follows.

Step 2: convergence of multivariate distributions.

We have to show that for any $n \geq 2$ and any $0 \leq t_1 < \cdots < t_n$ the weak convergence of finite-dimensional distributions holds

$$P_{t_1,\ldots,t_n}^{Z_a} \Rightarrow P_{t_1,\ldots,t_n}^{B^H}, \qquad \text{as } a\lambda \to \infty.$$

In other words,

$$(Z_a(t_1),\ldots,Z_a(t_n)) \Rightarrow (B^H(t_1),\ldots,B^H(t_n)) \qquad (13.25)$$

in $\mathbb{R}^n$.

Unlike in the case of Wiener process (Theorem 13.1), the limiting process B^H has dependent increments, hence we need a different strategy for proving (13.25). The one provided below actually gives an alternative proof for Theorem 13.1.

By Cramér–Wold Theorem 4.8 it is sufficient to prove that every univariate projection converges, i.e. for any real $c_1,\ldots,c_n$ we have

$$\sum_{j=1}^n c_j Z_a(t_j) \Rightarrow \sum_{j=1}^n c_j B^H(t_j). \qquad (13.26)$$

Recall that by (13.23) for each $j \leq n$ we have

$$Z_a(t_j) = K^{-1/2} \int_{\mathcal{R}} f_{t_j,a} d\widetilde{N},$$

where

$$f_{t,a}(s,u,r) := \frac{r\ell_{at}(s,u)}{a^{(3-\gamma)/2}\lambda^{1/2}}.$$

Then (13.26) writes as

$$S_a := K^{-1/2} \int_{\mathcal{R}} \left(\sum_{j=1}^n c_j f_{t_j,a} \right) d\widetilde{N} \Rightarrow \sum_{j=1}^n c_j B^H(t_j) := S.$$

As usual, the proof of convergence of centered Poisson integrals to a normal distribution is based on Proposition 8.1. First, we must check that

$$\mathbb{V}ar S_a \to \mathbb{V}ar S;$$

Indeed, by (13.22)

$$\mathbb{V}ar S_a = \mathbb{V}ar \sum_{j=1}^n c_j Z_a(t_j)$$

$$= \sum_{i,j=1}^n c_i c_j \mathbb{E}\, Z_a(t_i) Z_a(t_j)$$

$$\to \sum_{i,j=1}^n c_i c_j \mathbb{E}\, B^H(t_i) B^H(t_j)$$

$$= \mathbb{V}ar \sum_{j=1}^n c_j B^H(t_j) = \mathbb{V}ar S.$$

Now it remains to verify the second assumption (8.3) of Proposition 8.1, i.e. we check that for any $\varepsilon > 0$

$$\int_{\{|f_a| > \varepsilon\}} f_a^2 d\mu \to 0,$$

where

$$f_a = \sum_{j=1}^{n} c_j f_{t_j, a}.$$

Since the function $\ell_t(s, u)$ is increasing in argument t we have $f_{t,a} \leq f_{1,a}$ for any $t \leq 1$. It follows that

$$0 \leq |f_a| \leq \sum_{j=1}^{n} |c_j| f_{1,a} := C f_{1,a},$$

hence

$$\{|f_a| > \varepsilon\} \subset \{f_{1,a} > \varepsilon/C\}$$

and

$$\int_{\{|f_a| > \varepsilon\}} f_a^2 d\mu \leq C^2 \int_{\{f_{1,a} > \varepsilon/C\}} f_{1,a}^2 d\mu.$$

Yet by letting $t := 1$ in (13.24), we know that

$$\int_{\{f_{1,a} > \varepsilon\}} f_{1,a}^2 d\mu \to 0,$$

hence, (13.26) follows.

Step 3: weak convergence in $\mathbb{C}[0, 1]$. We will use Theorem 11.20 as a tool for establishing weak convergence. Since the finite-dimensional convergence is verified, it remains to check the moment inequality for increments (11.7). Since the workload has stationary increments, it is sufficient to prove that

$$\mathbb{E} |Z_a(t)|^a \leq C t^b \qquad \text{for all } t > 0,$$

with appropriate $a, C > 0$ and $b > 1$. In our case $a = 2$ is the most convenient choice. By using (12.18) and the definition of b we have

$$\mathbb{E} \, Z_a(t)^2 = \frac{\mathrm{Var} W^*(at)^2}{b^2} \leq \frac{\widehat{c_U} \lambda \, \mathbb{E} R^2 \, (at)^{3-\gamma}}{b^2}$$

$$= \frac{(2-\gamma)(\gamma-1)(3-\gamma)\widehat{c_U}(at)^{3-\gamma}}{2c_U a^{3-\gamma}} := C t^{3-\gamma},$$

and we are done, since $3 - \gamma > 1$. $\square$

We recommend [29, 85, 102] for exposition of many phenomena inherent to long range dependence.

Remark 13.5. There is another type of long range dependence limit theorems that we do not consider here. It concerns asymptotic behavior of nonlinear functionals like $\int_0^a G(W^\circ(t))\,dt$ as $a \to \infty$, see the seminal works [26, 86, 99, 100]. The limiting processes for this scheme are so called Rosenblatt-type processes. They are infinitely divisible but neither Gaussian nor stable and present an important example of multiple Wiener integrals [65, 79].

13.4 *Convergence to a stable Lévy process*

We consider here two cases where the main part of the workload is due to the service processes with duration shorter than the observation horizon (whence the independence of increments of the limiting process) and at the same time there are "short" service processes whose contribution $r\ell_{at}(s, u)$ to the workload $W^*(a)$ is comparable to the size of this workload (whence the positive jumps of the limiting process). Since we also expect that the limiting process should be self-similar, spectrally positive Lévy stable processes are the only reasonable candidates for describing the workload behavior in a long run.

We will prove two theorems about the convergence of the workload process to the stable process. Interestingly, the interpretation of the results will be somewhat different: in the first case, the limit is determined by the heavy tails of the distribution of required resources $\mathbb{P}(R \geq r)$ as $r \to \infty$; in the second case it is determined by the heavy tails of duration times $\mathbb{P}(U \geq u)$ as $u \to \infty$. In both cases the product UR belongs to the domain of attraction of a stable distribution, cf. (13.29), (13.40). Therefore, the spirit of Theorems 13.6 and 13.11 below is very much the same as that of classical Theorem 3.2.

Normalized workload process Z_a similar to a stable Lévy process is represented on Figure 13.5. Almost vertical segments of the graph correspond to important service processes that contribute substantially to the workload while being active on a relatively short time interval. Decreasing segments of a graph correspond to time intervals where negative centering of the process is dominating.

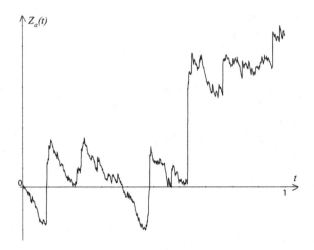

Fig. 13.5   Normalized workload process Z_a similar to a stable Lévy process.

13.4.1   *Dominating resource distribution*

Here we consider the zone of parameters $1 < \delta < \gamma \le 2$. By comparing (12.6) and (12.7), we see that R has the heavier distribution tails than U. These heavy tails of R lead to appearance of service processes requiring many resources and determine the limit behavior of the workload.

Notice that(12.6) yields for some $c > 0$ and all $u > 0$

$$\mathbb{P}(U > u) \le c\,u^{-\gamma}. \tag{13.27}$$

Similarly, (12.7) yields for some $c > 0$ and all $r > 0$

$$\mathbb{P}(R > r) \le c\,r^{-\delta}. \tag{13.28}$$

In the zone $\delta < \gamma$ we have $\mathbb{E}\,(U^\delta) < \infty$, since by (13.27) we have

$$\begin{aligned}
\mathbb{E}\,(U^\delta) &= \int_0^\infty \mathbb{P}(U^\delta > u)du \\
&= \int_0^\infty \mathbb{P}(U > u^{1/\delta})du \\
&\le 1 + c \int_1^\infty u^{-\gamma/\delta}du < \infty.
\end{aligned}$$

For the tail probabilities of RU under our basic assumptions (12.6) and (12.7) we have the following asymptotics,

$$\mathbb{P}\,(RU \ge y) \sim c_R \mathbb{E}\,(U^\delta)\,y^{-\delta}, \qquad \text{as} \quad y \to \infty. \tag{13.29}$$

Indeed, for any $\varepsilon > 0$ there exists a large M such that for all $r \geq M$ we have

$$\mathbb{P}(R > r) \leq \frac{c_R + \varepsilon}{r^\delta} .$$

Thus

$$\mathbb{P}(RU \geq y) \leq \mathbb{P}\left(U \geq \frac{y}{M}\right) + \int_0^{y/M} \frac{(c_R + \varepsilon)u^\delta}{y^\delta} F_U(du).$$

For large y this yields

$$\mathbb{P}(RU \geq y) \leq 2c_U(M/y)^\gamma + (c_R + \varepsilon)\mathbb{E}\, U^\delta y^{-\delta} \sim (c_R + \varepsilon)\mathbb{E}\,(U^\delta)y^{-\delta}.$$

Letting $\varepsilon \to 0$ yields the upper bound in (13.29). The lower bound follows by the same argument, with a simple dropping off the first term. Notice that we also have a uniform upper bound: for some $c > 0$ and all $y > 0$

$$\mathbb{P}(RU \geq y) \leq c\, y^{-\delta}. \qquad (13.30)$$

Now we are ready to state the first of two theorems treating convergence to a Lévy stable process. Let a constant B be defined by identity

$$B^\delta = c_R\, \delta\, \mathbb{E}\,(U^\delta)$$

and choose the scaling factor

$$b = b(a, \lambda) := B(a\lambda)^{1/\delta}. \qquad (13.31)$$

Theorem 13.6. *Let $\delta < \gamma \leq 2$. Assume that $a \to \infty, a\lambda \to \infty$. Define the scaling factor b by (13.31). Let $\mathcal{Y}(t), 0 \leq t \leq 1$, be a strictly δ-stable Lévy process such that $P_{\mathcal{Y}(1)} = \widetilde{\mathcal{S}}(0, 1, \delta)$. Then we have*

$$Z_a \overset{f.d.d.}{\longrightarrow} \mathcal{Y}.$$

Moreover,

$$Z_a \Rightarrow \mathcal{Y} \text{ in } (\mathbb{D}[0, 1], \rho_{_{\mathbb{D}, M}}).$$

Proof.

Step 1: contribution of the past is negligible. Write

$$Z_a(t) = D_a(t) + Z_a^+(t)$$
$$:= \int_{\mathcal{R}} \frac{r\ell_{at}(s, u)}{B(a\lambda)^{1/\delta}}\, 1_{\{s \leq 0\}} d\widetilde{N} + \int_{\mathcal{R}} \frac{r\ell_{at}(s, u)}{B(a\lambda)^{1/\delta}}\, 1_{\{0 < s \leq at\}} d\widetilde{N}.$$

We show that the contribution of the past $D_a(t)$ is asymptotically negligible, namely

$$D_a(t) \Rightarrow 0. \qquad (13.32)$$

For any $h > 0$ and $x > 0$ we have

$$\mu \left\{ \frac{r\ell_{at}(s,u)}{B(a\lambda)^{1/\delta}} \geq x; s \leq 0 \right\}$$

$$= \mu \left\{ \frac{r\ell_{at}(s,u)}{B(a\lambda)^{1/\delta}} \geq x; \begin{matrix} s \leq 0 \\ u \leq ha \end{matrix} \right\} + \mu \left\{ \frac{r\ell_{at}(s,u)}{B(a\lambda)^{1/\delta}} \geq x; \begin{matrix} s \leq 0 \\ u > ha \end{matrix} \right\}$$

$$\leq \mu \left\{ \frac{ru}{B(a\lambda)^{1/\delta}} \geq x; -ha \leq s \leq 0 \right\}$$

$$+\mu \left\{ \frac{ra}{B(a\lambda)^{1/\delta}} \geq x; -u \leq s \leq 0, u > ha \right\}$$

$$\leq ha\lambda \, \mathbb{P} \left\{ \frac{RU}{B(a\lambda)^{1/\delta}} \geq x \right\} + \mathbb{P} \left\{ \frac{Ra}{B(a\lambda)^{1/\delta}} \geq x \right\} \lambda \, \mathbb{E} \left(U \mathbf{1}_{\{U > ha\}} \right).$$

By using the known tail bounds (13.29), (12.7), (12.6), we obtain

$$\mathbb{P} \left\{ \frac{RU}{B(a\lambda)^{1/\delta}} \geq x \right\} \leq c_1 B^{-\delta} (a\lambda)^{-1} x^{-\delta},$$

$$\mathbb{P} \left\{ \frac{Ra}{B(a\lambda)^{1/\delta}} \geq x \right\} \leq c_2 B^{-\delta} (a\lambda)^{-1} a^\delta x^{-\delta},$$

$$\mathbb{E} \left(U \mathbf{1}_{\{U > ha\}} \right) \leq c_3 (ha)^{1-\gamma}$$

with appropriate c_1, c_2, c_3. It follows that

$$\mu \left\{ \frac{r\ell_{at}(s,u)}{B(a\lambda)^{1/\delta}} \geq x; s \leq 0 \right\} \leq c_4 \left(h + a^{\delta-\gamma} h^{1-\gamma} \right) x^{-\delta}.$$

By choosing order-optimal $h := a^{\frac{\delta}{\gamma}-1}$, we have

$$\mu \left\{ \frac{r\ell_{at}(s,u)}{B(a\lambda)^{1/\delta}} \geq x; s \leq 0 \right\} \leq 2c_4 a^{\frac{\delta}{\gamma}-1} x^{-\delta}.$$

Since $\delta < \gamma$ and $a \to \infty$, we have $a^{\frac{\delta}{\gamma}-1} \to 0$, and (13.32) follows, cf. Exercise 8.6.

Step 2: convergence of one-dimensional distributions. We must show that

$$Z_a(t) \Rightarrow \mathcal{Y}(t),$$

which means $P_{Z_a(t)} \Rightarrow \widetilde{\mathcal{S}}(0, t, \delta)$. In view of (13.32) it is sufficient to prove that

$$P_{Z_a^+(t)} \Rightarrow \widetilde{\mathcal{S}}(0, t, \delta),$$

by using Corollary 8.5 of Proposition 8.4.

Recall that

$$Z_a^+(t) = \int_{\mathcal{R}} \frac{r\ell_{at}(s,u)}{B(a\lambda)^{1/\delta}} \mathbf{1}_{\{0<s\leq at\}} d\widetilde{N},$$

where the kernel $\ell_{at}(s,u)$ is defined in (12.4).

We have to check the limiting relations (8.4) and (8.5), as well as the uniform bound (8.8). Since the integrands are positive, the relation (8.5) with $c_- = 0$ is obvious (no negative jumps). Next, we concentrate our efforts on the verification of (8.4).

For each $x > 0$ we have to check that

$$\mu\left\{(s,u,r): \frac{r\ell_{at}(s,u)}{B(a\lambda)^{1/\delta}} \geq x; s > 0\right\} \to \frac{t}{\delta x^\delta}. \tag{13.33}$$

Let us start with a lower bound for the measure we are interested in. The idea is to use that $\ell_{at}(s,u) = u$ (i.e. the service process entirely fits within the observation interval) on a sufficiently large set. Fix a large $M > 0$. Clearly,

$$\mu\left\{(s,u,r): \frac{r\ell_{at}(s,u)}{B(a\lambda)^{1/\delta}} \geq x; s > 0\right\}$$

$$\geq \mu\left\{(s,u,r): \frac{r\,u}{B(a\lambda)^{1/\delta}} \geq x, s \in (0, at-M], u \leq M\right\}$$

$$\geq \lambda \cdot (at - M) \int_0^M \mathbb{P}\left\{R \geq \frac{xB(a\lambda)^{1/\delta}}{u}\right\} F_U(du).$$

Since $a\lambda \to \infty$ and u is bounded from above in the integration domain, we may apply (12.7) and get

$$\int_0^M \mathbb{P}\left\{R \geq \frac{xB(a\lambda)^{1/\delta}}{u}\right\} F_U(du)$$

$$\sim c_R x^{-\delta} B^{-\delta} (a\lambda)^{-1} \int_0^M u^\delta F_U(du)$$

$$= c_R x^{-\delta} B^{-\delta} (a\lambda)^{-1} \mathbb{E}\left(U^\delta \mathbf{1}_{\{U \leq M\}}\right).$$

By using the definition of B we arrive at

$$\liminf \mu\left\{(s,u,r): \frac{r\ell_{at}(s,u)}{B(a\lambda)^{1/\delta}} \geq x; s \geq 0\right\} \geq \frac{t}{\delta x^\delta} \frac{\mathbb{E}\left(U^\delta \mathbf{1}_{\{U \leq M\}}\right)}{\mathbb{E}\left(U^\delta\right)}.$$

By letting $M \to \infty$, we obtain the desired lower bound in (13.33),

$$\liminf \mu\left\{(s,u,r): \frac{r\ell_{at}(s,u)}{B(a\lambda)^{1/\delta}} \geq x; s \geq 0\right\} \geq \frac{t}{\delta x^\delta}.$$

Moving towards the upper bound, we use the bound $\ell_{at}(s,u) \leq u$ and obtain

$$\mu\left\{(s,u,r): \frac{r\ell_{at}(s,u)}{B(a\lambda)^{1/\delta}} \geq x, 0 < s \leq at\right\}$$

$$\leq \mu\left\{(s,u,r): \; ru \geq xB(a\lambda)^{1/\delta}, s \in (0,at]\right\}$$

$$= a\lambda t \; \mathbb{P}\left\{RU \geq xB(a\lambda)^{1/\delta},\right\}$$

$$\sim a\lambda t \; c_R E(U^\delta)\left(xB(a\lambda)^{1/\delta}\right)^{-\delta} = t\delta^{-1}x^{-\delta},$$

where we used asymptotics (13.29) and the definition of B.

Therefore, the upper bound in (13.33) is proved.

The uniform bound (8.8) follows in the same way. Namely, we will show that

$$\mu\left\{(s,u,r): \frac{r\,\ell_{at}(s,u)}{b} \geq x\right\} \leq \frac{Ct^{\delta-1}}{x^\delta} \qquad (13.34)$$

with appropriate $C > 0$ for all $a \geq 1, x > 0, t \in [0,1]$. Let us split the domain in two, according to the inequalities $u \leq at$ or $u > at$. In the first domain we use $\ell_{at}(s,u) \leq u$ and $s \in [-u,at] \subset [-at,at]$ and the uniform tail bound (13.30) to obtain

$$\mu\left\{(s,u,r): \frac{r\,\ell_{at}(s,u)}{b} \geq x, u \leq at\right\}$$

$$\leq \mu\{(s,u,r): s \in [-at,at], ru \geq Bx(\lambda a)^{1/\delta}\}$$

$$= 2\lambda\, at\, \mathbb{P}\{RU \geq Bx(\lambda a)^{1/\delta}\}$$

$$\leq 2\lambda\, at\, C[Bx(\lambda a)^{1/\delta}]^{-\delta} = 2CB^{-\delta}\, t\, x^{-\delta}$$

$$\leq 2CB^{-\delta}\, t^{1-\delta}\, x^{-\delta}.$$

In the second domain we use $\ell_{at}(s,u) \leq u$ and $s \in [-u,at] \subset [-u,u]$ and the uniform tail bound (13.28) to obtain

$$\mu\left\{(s,u,r): \frac{r\,\ell_{at}(s,u)}{b} \geq x, u > at\right\}$$

$$\leq \mu\{(s,u,r): s \in [-u,u], rat \geq Bx(\lambda a)^{1/\delta}\}$$

$$= 2\lambda\, \mathbb{E}\left(U\mathbf{1}_{\{U>at\}}\right) \mathbb{P}\left\{R \geq \frac{Bx(\lambda a)^{1/\delta}}{at}\right\}$$

$$\leq 2\lambda C_1(at)^{1-\gamma}C_2\left[\frac{Bx(\lambda a)^{1/\delta}}{at}\right]^{-\delta}$$

$$= 2C_1C_2B^{-\delta}a^{\delta-\gamma}t^{1+\delta-\gamma}x^{-\delta}.$$

Since $a \geq 1$ and $\delta < \gamma$, we may drop the factor $a^{\delta-\gamma}$. Since $\gamma < 2$ and $t \leq 1$, we have $t^{1+\delta-\gamma} \leq t^{\delta-1}$. Therefore,

$$\mu\left\{(s,u,r): \frac{r\,\ell_{at}(s,u)}{b} \geq x, u > at\right\} \leq 2C_1C_2B^{-\delta}t^{\delta-1}x^{-\delta},$$

and we obtain (13.34) by adding up estimates for both parts.

Now Corollary 8.5 applies and yields the desired convergence.

Step 3: convergence of distributions for increments.

We just repeat the arguments from the finite variance case (Theorem 13.1): Recall that Z_a inherits from W^* the property to have stationary increments. This means that for any non-negative $t_1 \leq t_2$, the increment $Z_a(t_2) - Z_a(t_1)$ is equidistributed with $Z_a(t_2 - t_1)$. Therefore, we may restate the result of Step 2 as

$$Z_a(t_2) - Z_a(t_1) \Rightarrow \mathcal{Y}(t_2) - \mathcal{Y}(t_1), \qquad (13.35)$$

or $P_{Z_a(t_2)-Z_a(t_1)} \Rightarrow \tilde{\mathcal{S}}(0, t_2 - t_1, \delta)$.

Step 4: convergence of multivariate distributions.

We have to show that for any $n \geq 2$ and any $0 \leq t_1 < \cdots < t_n$ the weak convergence of finite-dimensional distributions holds

$$P^{Z_a}_{t_1,\ldots,t_n} \Rightarrow P^{\mathcal{Y}}_{t_1,\ldots,t_n}, \qquad \text{as } a\lambda \to \infty.$$

Again, the arguments developed in finite variance case apply. We pass to the differences (13.11) and notice that (13.35) provides the component-wise convergence

$$(\Delta_a)_j \Rightarrow (\Delta_{\mathcal{Y}})_j, \qquad j = 1, 2, \ldots, n.$$

Then, extracting independent parts, we split the increment Δ_a, as done in (13.13), (13.15). The components $(\Delta'_a)_j$ of the main part are independent, and it remains to establish that we have $(\Delta''_a)_j \Rightarrow 0$. By stationarity, it is sufficient to consider only $j = 1$. Finally, for $j = 1$ the claim $(\Delta''_a)_1 \Rightarrow 0$ coincides with (13.32) that has been already verified.

The remainder of the proof of multivariate convergence is a verbatim repetition of that for the finite variance case (Theorem 13.1).

Step 5: weak convergence in $(\mathbb{D}[0,1], \rho_{\mathbb{D},M})$.

In order to establish weak convergence of Z_a in the space $(\mathbb{D}[0,1], \rho_{\mathbb{D},M})$, we have to verify condition (11.10) of Theorem 11.22. Since Z_a has stationary increments, there is no loss of generality to consider a time interval starting at zero. We fix two instants $0 \leq s \leq t$ and begin evaluation with the bounds

$$\mathbb{P}\left(\min_{x \in [0, Z_a(t)]} |Z_a(s) - x| \geq \varepsilon\right)$$
$$= \mathbb{P}\left(Z_a(s) \geq \max\{0, Z_a(t)\} + \varepsilon\right) + \mathbb{P}\left(Z_a(s) \leq \min\{0, Z_a(t)\} - \varepsilon\right)$$
$$\leq \mathbb{P}\left(Z_a(s) \geq Z_a(t) + \varepsilon\right) + \mathbb{P}\left(Z_a(s) \leq -\varepsilon\right)$$
$$= \mathbb{P}\left(Z_a(t-s) \leq -\varepsilon\right) + \mathbb{P}\left(Z_a(s) \leq -\varepsilon\right). \tag{13.36}$$

It remains to evaluate the probability of a highly unlikely event: fast negative drift on a short time interval for a process having no negative jumps. This will be done by using the following estimate.

Lemma 13.7. *Let*

$$I := \int_{\mathcal{R}} f d\widetilde{N}$$

be a centered Poisson integral of a non-negative function f over a measure space $\mathcal{R}, \mu$. Assume that for some $C > 0, \alpha \in (1,2)$ and any $u > 0$ we have

$$\mu\{x \in \mathcal{R} : f(x) \geq u\} \leq \frac{C}{u^\alpha}.$$

Then for any $\varepsilon > 0$ it is true that

$$\mathbb{P}(I \leq -\varepsilon) \leq \exp\left\{-A(\alpha)\left(\varepsilon^\alpha / C\right)^{\frac{1}{\alpha-1}}\right\}, \tag{13.37}$$

where the constant $A(\alpha)$ depends only on α.

Proof.   Since we aim to evaluate deviation probability through exponential Chebyshev inequality, our first task is evaluation of exponential moments, this time with negative argument. Using variable change and integration by parts, for any $\tau \geq 0$ we easily find

$$\mathbb{E} \exp\{-\tau I\} = \exp\left\{\int_{\mathcal{R}} \left(e^{-\tau f(x)} - 1 + \tau f(x)\right) \mu(dx)\right\}$$
$$= \exp\left\{\int_0^\infty \left(e^{-\tau u} - 1 + \tau u\right) d\mu\{x : f(x) \leq u\}\right\}$$
$$= \exp\left\{\tau \int_0^\infty \left(1 - e^{-\tau u}\right) \mu\{x : f(x) \geq u\} du\right\}$$
$$\leq \exp\left\{\tau \int_0^\infty \left(1 - e^{-\tau u}\right) \frac{C \, du}{u^\alpha}\right\}$$
$$= \exp\left\{C\tau^\alpha \int_0^\infty \left(1 - e^{-v}\right) \frac{dv}{v^\alpha}\right\}$$
$$:= \exp\left\{CA_1(\alpha)\tau^\alpha\right\}.$$

By exponential Chebyshev inequality,

$$\mathbb{P}(I \le -\varepsilon) \le \frac{\mathbb{E} \exp\{-\tau I\}}{\exp\{\tau\varepsilon\}} \le \exp\{CA_1(\alpha)\tau^\alpha - \tau\varepsilon\}.$$

By plugging in the optimal parameter value

$$\tau := \left(\frac{\varepsilon}{\alpha A_1(\alpha)C}\right)^{\frac{1}{\alpha-1}}$$

we arrive at (13.37) with

$$A(\alpha) := \frac{\alpha - 1}{(A_1(\alpha)\alpha^\alpha)^{\frac{1}{\alpha-1}}} \ .$$

$\square$

Back to our case,

$$I := Z_a(t) = \int_\mathcal{R} \frac{r\,\ell_{at}(s,u)}{b}\,d\tilde{N},$$

assume that for some $C, p > 0$ we have an integrand bound

$$\mu\left\{(s,u,r) : \frac{r\,\ell_{at}(s,u)}{b} \ge x\right\} \le \frac{Ct^p}{x^\alpha} \tag{13.38}$$

for all $a, x > 0, t \in [0,1]$. Then (13.37) yields

$$\mathbb{P}(Z_a(t) \le -\varepsilon) \le \exp\left\{-A(\alpha)\,(\varepsilon^\alpha/Ct^p)^{\frac{1}{\alpha-1}}\right\}.$$

By using a bound

$$\exp\left\{-A(\alpha)y^{\frac{1}{\alpha-1}}\right\} \le L(\alpha)y^{-2/p}, \qquad y > 0,$$

with appropriate constant $L(\alpha)$, we obtain

$$\mathbb{P}(Z_a(t) \le -\varepsilon) \le L(\alpha)C^{2/p}\varepsilon^{-2\alpha/p}t^2.$$

We obtain now from (13.36)

$$\mathbb{P}\left(\min_{x\in[0,Z_a(t)]}|Z_a(s) - x| \ge \varepsilon\right) \le 2L(\alpha)C^{2/p}\varepsilon^{-2\alpha/p}t^2,$$

as required in (11.10).

It remains to recall that (13.34) provides the bound precisely of the form (13.38) with $\alpha = \delta$ and $p = \delta - 1$.

$\square$

Remark 13.8. We proved the weak convergence of Z_a to a Lévy process $\mathcal{Y}$ in the space $(\mathbb{D}[0,1], \rho_{\mathbb{D},M})$. Let us stress that there is no chance to prove the weak convergence in the space $(\mathbb{D}[0,1], \rho_{\mathbb{D},J})$ (which would be a better result) because $\mathbb{P}(Z_a \in \mathbb{C}[0,1]) = 1$ but $\mathbb{P}(\mathcal{Y} \in \mathbb{C}[0,1]) = 0$, and $\mathbb{C}[0,1]$ is a closed set in $(\mathbb{D}[0,1], \rho_{\mathbb{D},J})$.

Exercise 13.9. ([2]) Let us drop the assumption $\delta > 1$ and consider the case $\delta < 1 < \gamma$. Then the expectation $\mathbb{E} R$, hence $\mathbb{E} W^*(t)$ become infinite, and we must renounce from centering in the definition of the process $Z_a(\cdot)$. Thus let $Z_a(t) := \frac{W^*(at)}{b}$, $t \in [0,1]$. With this modification, prove a complete analogue of Theorem 13.6.

Exercise 13.10. Consider the case $\gamma = \delta$ corresponding to the boundary diagonal on Figure 13.1. Prove first the asymptotics

$$\mathbb{P}(RU > y) \sim c_R c_U \, \delta \, y^{-\delta} \ln y, \qquad \text{as} \quad y \to \infty,$$

as a counterpart to (13.29). Prove an analogue to Theorem 13.6 with scaling factor

$$b(a, \lambda) := B \left(a\lambda \ln(a\lambda) \right)^{1/\delta},$$

where $B^\delta := c_R c_U \delta$.

13.4.2 *Dominating service duration*

Here we consider the zone of parameters $1 < \gamma < \delta \leq 2$. By comparing (12.7) and (12.6), we see that this time U has the heavier distribution tails than R. These heavy tails of U lead to appearance of sufficiently long service processes determine the limit behavior of the workload.

However, these long service process should not be too long to destroy independence of increments. This requirement is embodied in the following *low intensity condition*:

$$\frac{\lambda}{a^{\gamma-1}} \to 0, \tag{13.39}$$

which is opposite to the high intensity condition (13.19) and means that the number of service processes of duration comparable to observation horizon a tends to zero.

Next, we pass to the scaling arrangement. First of all, notice that in this zone $\mathbb{E}(R^\gamma) < \infty$, since by (13.28) we obtain

$$\mathbb{E}(R^\gamma) = \int_0^\infty \mathbb{P}(R^\gamma > r)dr$$

$$= \int_0^\infty \mathbb{P}(R > r^{1/\gamma})dr$$

$$\leq 1 + c \int_1^\infty r^{-\delta/\gamma}dr < \infty.$$

Similarly to (13.29), this time we have

$$\mathbb{P}(RU \geq y) = \int \mathbb{P}(U \geq \tfrac{y}{r})F_R(dr)$$

$$\sim \int \frac{c_U r^\gamma}{y^\gamma}F_R(dr)$$

$$= \frac{c_U \mathbb{E}(R^\gamma)}{y^\gamma}, \quad \text{as} \quad y \to \infty. \tag{13.40}$$

Notice that we also have a uniform upper bound: for some $c > 0$ and all $y > 0$ it is true that

$$\mathbb{P}(RU \geq y) \leq c\, y^{-\gamma}. \tag{13.41}$$

Now we are ready to state the second theorem treating convergence to a Lévy stable process. Let a constant B be defined by identity

$$B^\gamma = c_U\, \gamma\, \mathbb{E}(R^\gamma)$$

and choose the scaling factor

$$b = b(a, \lambda) := B(a\lambda)^{1/\gamma}. \tag{13.42}$$

Theorem 13.11. *Let $\gamma < \delta \leq 2$. Assume that $a \to \infty$, $a\lambda \to \infty$, and that low intensity condition (13.39) holds. Define the scaling factor b by (13.42).*

Let $\mathcal{Y}(t), 0 \leq t \leq 1$, be a strictly γ-stable Lévy process such that $P_{\mathcal{Y}(1)} = \widetilde{\mathcal{S}}(0, 1, \gamma)$. Then we have

$$Z_a \xrightarrow{f.d.d.} \mathcal{Y}.$$

Moreover,

$$Z_a \Rightarrow \mathcal{Y} \text{ in } (\mathbb{D}[0, 1], \rho_{\mathbb{D}, M}).$$

Proof.

Step 1: contribution of the past is negligible. Write

$$Z_a(t) = D_a(t) + Z_a^+(t)$$

$$:= \int_{\mathcal{R}} \frac{r\ell_{at}(s, u)}{B(a\lambda)^{1/\gamma}} \mathbf{1}_{\{s \leq 0\}} d\widetilde{N} + \int_{\mathcal{R}} \frac{r\ell_{at}(s, u)}{B(a\lambda)^{1/\gamma}} \mathbf{1}_{\{0 < s \leq at\}} d\widetilde{N}.$$

We show that the contribution of the past $D_a(t)$ is asymptotically negligible, namely,

$$D_a(t) \Rightarrow 0. \tag{13.43}$$

Let us split this variable into a sum according to the length of duration u,

$$D_a := (D_{a,1} - \mathbb{E} D_{a,1}) + D_{a,2},$$

where

$$D_{a,1} := \int_{\mathcal{R}} \frac{r\ell_{at}(s,u)}{B(a\lambda)^{1/\gamma}} \, \mathbf{1}_{\{u > ha, s \le 0\}} \, dN,$$

$$D_{a,2} := \int_{\mathcal{R}} \frac{r\ell_{at}(s,u)}{B(a\lambda)^{1/\gamma}} \, \mathbf{1}_{\{u \le ha, s \le 0\}} \, d\widetilde{N},$$

and a threshold $h > 0$ will be specified later. Let $\theta := \frac{\lambda}{a^{\gamma-1}}$. We have

$$
\begin{aligned}
\mathbb{P}\{D_{a,1} \ne 0\} &\le \mu\{u > ha, s \in [-u, 0]\} \\
&= \lambda \mathbb{E}\left(U \mathbf{1}_{\{U > ha\}}\right) \le c_1 \lambda (ha)^{1-\gamma} \\
&= c_1 \theta h^{1-\gamma};
\end{aligned}
$$

$$
\begin{aligned}
\mathbb{E} D_{a,1} &\le \int_{\mathcal{R}} \frac{ra}{B(a\lambda)^{1/\gamma}} \, \mathbf{1}_{\{u > ha, s \in [-u, 0]\}} \, d\mu, \\
&= \frac{\mathbb{E}R\, a}{B(a\lambda)^{1/\gamma}} \lambda \mathbb{E}\left(U \mathbf{1}_{\{U > ha\}}\right) \\
&\le \frac{\mathbb{E}R\, a}{B(a\lambda)^{1/\gamma}} c_1 \lambda (ha)^{1-\gamma} \\
&:= c_2 \theta^{1-1/\gamma} h^{1-\gamma}.
\end{aligned}
$$

By choosing $h := \theta^{1/2\gamma}$ and using (13.39) we have

$$
\begin{aligned}
\mathbb{P}\{D_{a,1} \ne 0\} &\le c_1 \theta^{(1+\gamma)/2\gamma} \to 0; \\
0 \le \mathbb{E} D_{a,1} &\le c_2 \theta^{(\gamma-1)/2\gamma} \to 0.
\end{aligned}
$$

Hence $D_{a,1} - \mathbb{E} D_{a,1} \Rightarrow 0$. By considering the integrand of $D_{a,2}$, we see that

$$
\begin{aligned}
\mu\left\{ \frac{r\ell_{at}(s,u)}{B(a\lambda)^{1/\gamma}} \ge x; u \le ha; s \le 0 \le s + u \right\} \\
\le \mu\left\{ ru \ge B(a\lambda)^{1/\gamma} x; s \in [-ha, 0] \right\} \\
= \mathbb{P}\left\{ RU \ge B(a\lambda)^{1/\gamma} x \right\} \lambda ha \\
\le c_3 (a\lambda)^{-1} x^{-\gamma} \lambda ha = c_3 h x^{-\gamma}.
\end{aligned}
$$

Since by (13.39) $h := \theta^{1/2\gamma} \to 0$, it follows from Exercise 8.6 that $D_{a,2} \Rightarrow 0$.

Step 2: convergence of one-dimensional distributions.
We must show that

$$Z_a(t) \Rightarrow \mathcal{Y}(t),$$

or $P_{Z_a(t)} \Rightarrow \tilde{\mathcal{S}}(0, t, \gamma)$. In view of (13.43) it is sufficient to prove that $P_{Z_a^+(t)} \Rightarrow \tilde{\mathcal{S}}(0, t, \gamma)$, by using Corollary 8.5 of Proposition 8.4.
Recall that

$$Z_a^+(t) = \int_{\mathcal{R}} \frac{r\ell_{at}(s, u)}{B(a\lambda)^{1/\gamma}} \, 1_{\{0 < s \leq at\}} d\tilde{N},$$

where the kernel $\ell_{at}(s, u)$ is defined in (12.4).

We have to check the limiting relations (8.4) and (8.5), as well as the uniform bound (8.8). Since the integrands are positive, the relation (8.5) with $c_- = 0$ is obvious (no negative jumps). Next, we concentrate our efforts on the verification of (8.4).

For each $x > 0$ we have to check that

$$\mu\left\{(s, u, r) : \frac{r\ell_{at}(s, u)}{B(a\lambda)^{1/\gamma}} \geq x; 0 < s \leq at\right\} \to \frac{t}{\gamma x^\gamma}. \qquad (13.44)$$

Let us start with a lower bound for the measure we are interested in. The idea is to use that $\ell_{at}(s, u) = u$ (i.e. the service process entirely fits within the observation interval) on a sufficiently large set.

Taking (13.40) into account, for any $h \in (0, t)$ we obtain

$$\mu\left\{(s, u, r) : \frac{r\ell_{at}(s, u)}{B(a\lambda)^{1/\gamma}} \geq x, 0 < s \leq at\right\}$$

$$\geq \mu\left\{(s, u, r) : \frac{r\,u}{B(a\lambda)^{1/\gamma}} \geq x, s \in (0, a(t - h)], u \leq ha\right\}$$

$$\geq \mathbb{P}\left\{\frac{RU}{B(a\lambda)^{1/\gamma}} \geq x, U \leq ha\right\} \cdot \lambda \cdot (t - h)a.$$

Fix two constants for a while: a small one r_0 and a large one M, and notice that in virtue of (13.39) it is true that

$$\frac{(a\lambda)^{1/\gamma}}{r} \ll ha$$

uniformly on $r \geq r_0$. Therefore,

$$\mathbb{P}\left\{\frac{RU}{B(a\lambda)^{1/\gamma}} \geq x, U \leq ha\right\}$$

$$= \int \mathbb{P}\left\{\frac{xB(a\lambda)^{1/\gamma}}{r} \leq U \leq ha\right\} F_R(dr)$$

$$\geq \int_{r_0}^{\infty} \mathbb{P}\left\{\frac{xB(a\lambda)^{1/\gamma}}{r} \leq U \leq \frac{MxB(a\lambda)^{1/\gamma}}{r}\right\} F_R(dr)$$

$$\sim \int_{r_0}^{\infty} (1 - M^{-\gamma})c_U \left(\frac{xB(a\lambda)^{1/\gamma}}{r}\right)^{-\gamma} F_R(dr)$$

$$= \mathbb{E}\left(1_{\{R \geq r_0\}} R^{\gamma}\right)(1 - M^{-\gamma})\frac{c_U}{B^{\gamma}} x^{-\gamma}(a\lambda)^{-1}.$$

By combining this fact with the previous estimate and taking limits in r_0, M, h, we obtain the asymptotic bound $\mathbb{E}(R^{\gamma}) \frac{c_U}{B^{\gamma}} t x^{-\gamma}$. It remains to use the definition of the constant B, and we obtain the desired lower bound in (13.44),

$$\liminf \mu\left\{(s, u, r) : \frac{r\ell_{at}(s, u)}{B(a\lambda)^{1/\gamma}} \geq x; 0 < s \leq at\right\} \geq \frac{t}{\gamma x^{\gamma}}.$$

Moving towards the upper bound in (13.44), we use $\ell_{at}(s, u) \leq u$ and obtain

$$\mu\left\{(s, u, r) : \frac{r\ell_{at}(s, u)}{B(a\lambda)^{1/\gamma}} \geq x, 0 < s \leq at\right\}$$

$$\leq \mu\left\{(s, u, r) : ru \geq xB(a\lambda)^{1/\gamma}, s \in (0, at]\right\}$$

$$= a\lambda t \, \mathbb{P}\left\{RU \geq xB(a\lambda)^{1/\gamma}\right\}$$

$$\sim a\lambda t \, c_U \mathbb{E}(R^{\gamma}) \left(xB(a\lambda)^{1/\gamma}\right)^{-\gamma} = t\gamma^{-1}x^{-\gamma},$$

where we used asymptotics (13.40) and the definition of B.

Therefore, the upper bound in (13.44) is proved.

The uniform bound (8.8) follows by the same way. Namely, we will show that

$$\mu\left\{(s, u, r) : \frac{r\,\ell_{at}(s, u)}{b} \geq x\right\} \leq \frac{Ct}{x^{\gamma}} \tag{13.45}$$

with appropriate $C > 0$ for all $a, x > 0, t \in [0, 1]$. Let us split the domain in two, according to the inequalities $u \leq at$ or $u > at$. In the first domain

we use $\ell_{at}(s, u) \leq u$ and $s \in [-u, at] \subset [-at, at]$, as well as the tail bound
(13.41) to obtain

$$\mu\left\{(s, u, r) : \frac{r \, \ell_{at}(s, u)}{b} \geq x, u \leq at\right\}$$

$$\leq \mu\{(s, u, r) : s \in [-at, at], ru \geq Bx(\lambda a)^{1/\gamma}\}$$
$$= 2\lambda \, at \, \mathbb{P}\{RU \geq Bx(\lambda a)^{1/\gamma}\}$$
$$\leq 2\lambda \, at \, C[Bx(\lambda a)^{1/\gamma}]^{-\gamma} = 2CB^{-\gamma} t \, x^{-\gamma}.$$

In the second domain we use $\ell_{at}(s, u) \leq u$ and $s \in [-u, at] \subset [-u, u]$ to
obtain

$$\mu\left\{(s, u, r) : \frac{r \, \ell_{at}(s, u)}{b} \geq x, u > at\right\}$$

$$\leq \mu\{(s, u, r) : s \in [-u, u], rat \geq Bx(\lambda a)^{1/\delta}\}$$
$$= 2\lambda \, \mathbb{E}\left(U1_{\{U>at\}}\right) \mathbb{P}\left\{R \geq \frac{Bx(\lambda a)^{1/\gamma}}{at}\right\}$$
$$\leq 2\lambda C_1 (at)^{1-\gamma} \mathbb{P}\left\{R \geq \frac{Bx(\lambda a)^{1/\gamma}}{at}\right\}.$$

At this place we proceed differently for x satisfying $\frac{Bx(\lambda a)^{1/\gamma}}{at} \leq 1$ and
$\frac{Bx(\lambda a)^{1/\gamma}}{at} > 1$. In the first case we have

$$x^\gamma \leq \frac{(at)^\gamma}{B^\gamma(\lambda a)},$$

so we just drop the probability related to R in the bound and obtain

$$\mu\left\{(s, u, r) : \frac{r \, \ell_{at}(s, u)}{b} \geq x, u > at\right\} \leq 2\lambda C_1 (at)^{1-\gamma} = 2\lambda C_1 (at)^{1-\gamma} x^\gamma \, x^{-\gamma}$$

$$\leq 2\lambda C_1 (at)^{1-\gamma} \frac{(at)^\gamma}{B^\gamma(\lambda a)} x^{-\gamma}$$

$$= 2C_1 B^{-\gamma} t \, x^{-\gamma}.$$

In the second case we have

$$x \geq \frac{at}{B(\lambda a)^{1/\gamma}} ;$$

so we use the uniform tail estimate for R (13.28) and again obtain

$$\mu\left\{(s,u,r): \frac{r\,\ell_{at}(s,u)}{b} \geq x, u > at\right\}$$

$$\leq 2\lambda C_1 (at)^{1-\gamma} C_2 \left[\frac{Bx(\lambda a)^{1/\gamma}}{at}\right]^{-\delta}$$

$$= 2\lambda C_1 (at)^{1-\gamma} C_2 \left[\frac{B(\lambda a)^{1/\gamma}}{at}\right]^{-\delta} x^{-(\delta-\gamma)} x^{-\gamma}$$

$$\leq 2\lambda C_1 (at)^{1-\gamma} C_2 \left[\frac{B(\lambda a)^{1/\gamma}}{at}\right]^{-\delta} \left[\frac{at}{B(\lambda a)^{1/\gamma}}\right]^{-(\delta-\gamma)} x^{-\gamma}$$

$$= 2C_1 C_2 B^{-\gamma} t\, x^{-\gamma}.$$

Finally, we obtain (13.45) by adding up estimates for all parts.

Now Corollary 8.5 applies and yields the desired convergence $P_{Z_a^+(t)} \Rightarrow \widetilde{\mathcal{S}}(0,t,\gamma)$.

Convergence of finite-dimensional distributions and weak convergence in the space $(\mathbb{D}[0,1], \rho_{\mathbb{D},M})$ are justified by the same arguments as in Theorem 13.6.   □

Remark 13.12. As before, we have the weak convergence of Z_a to a Lévy process $\mathcal{Y}$ in the space $(\mathbb{D}[0,1], \rho_{\mathbb{D},M})$ but there is no weak convergence in the space $(\mathbb{D}[0,1], \rho_{\mathbb{D},J})$ because $\mathbb{P}(Z_a \in \mathbb{C}[0,1]) = 1$ but $\mathbb{P}(\mathcal{Y} \in \mathbb{C}[0,1]) = 0$, and $\mathbb{C}[0,1]$ is a closed set in $(\mathbb{D}[0,1], \rho_{\mathbb{D},J})$.

13.5   Convergence to Telecom processes

In this section we consider original limit theorems from [49] such that the limiting processes do not belong to any class widely known in the theory of limit theorems. They just can be written as integrals with respect to a stable or to a Poisson random measure. In [49] such processes are called *Telecom processes* – according to the application domain they come from.

13.5.1   Convergence to a stable Telecom process

We will prove here another limit theorem for the workload Z_a in the case $1 < \gamma < \delta < 2$. Recall that we already treated this case in Theorem 13.11 and showed that under *low* intensity condition the workload converges to a γ-stable Lévy process. Now we turn to the *high* intensity case. Here the limit is different because long service processes are not negligible anymore. Note the related difference of diagrams at Figures 13.1-13.2, zones C vs. E.

Let us choose a scaling factor as

$$b = Ba^{(\delta+1-\gamma)/\delta}\lambda^{1/\delta}, \tag{13.46}$$

where the constant B is defined by $B^\delta := c_R c_U \delta\gamma$.

Theorem 13.13. *Assume that $1 < \gamma < \delta < 2$, $a \to \infty$, and that high intensity condition (13.19) holds. Then with the scaling (13.46) it is true that*

$$Z_a \xrightarrow{f.d.d.} \mathcal{Z}_{\gamma,\delta},$$

where the process $\mathcal{Z}_{\gamma,\delta}(t)$ (called stable Telecom process) admits an integral representation

$$\mathcal{Z}_{\gamma,\delta}(t) = \int\int \ell_t(s, u) X(ds, du).$$

Here $\ell_t(s, u)$ is the kernel defined in (12.4), and X is a δ-stable independently scattered random measure with intensity $u^{-\gamma-1} ds du$, corresponding to the spectrally positive strictly δ-stable distribution $\widetilde{S}(0, 1, \delta)$.

Remark 13.14. The dependence $t \to \sigma_t$ defined below is not linear. This means that Telecom process $\mathcal{Z}_{\gamma,\delta}(t)$ is not a stable Lévy process.

Proof. We only prove the convergence of one-dimensional distributions. Recall that

$$Z_a(t) = \int_{\mathbb{R}} \frac{r\ell_{at}(s, u)}{Ba^{(\delta+1-\gamma)/\delta}\lambda^{1/\delta}} \, d\widetilde{N}.$$

On the other hand, by (7.34) the variable $\mathcal{Z}_{\gamma,\delta}(t)$ has the stable distribution $\widetilde{S}(0, \sigma_t, \delta)$, where parameter σ_t can be found by the formula

$$\sigma_t = \sigma_t(\gamma, \delta) = \|\ell_t\|_\delta^\delta = \int\int \ell_t(s, u)^\delta \frac{ds du}{u^{\gamma+1}}.$$

First of all, let us check that this integral is finite. Indeed, by splitting it into in two parts and using $\delta > \gamma$, we obtain

$$\int_{\{u \leq t\}} \int \ell_t(s,u)^\delta \frac{dsdu}{u^{\gamma+1}} \leq \int_0^t \int_{-u}^t ds \frac{u^\delta du}{u^{1+\gamma}}$$

$$= \int_0^t \frac{(u+t)u^\delta du}{u^{1+\gamma}}$$

$$\leq 2t \int_0^t \frac{u^\delta du}{u^{1+\gamma}} < \infty,$$

$$\int_{\{u \leq t\}} \int \ell_t(s,u)^\delta \frac{dsdu}{u^{\gamma+1}} \leq t^\delta \int_t^\infty \int_{-u}^t ds \frac{du}{u^{\gamma+1}}$$

$$= t^\delta \int_t^\infty \frac{(u+t)du}{u^{\gamma+1}}$$

$$\leq 2t^\delta \int_t^\infty \frac{du}{u^\gamma} < \infty.$$

Therefore, we may check the convergence $Z_a(t) \Rightarrow \mathcal{Z}_{\gamma,\delta}(t)$ by using sufficient conditions for convergence of Poisson integrals' distributions to a stable distribution.

We will use Corollary 8.5 of Proposition 8.4. We have to check the limiting relations (8.4) and (8.5), as well as the uniform bound (8.8). Since the integrands are positive, the relation (8.5) is obvious. Now we pass to the verification of (8.4). We have to check that for any $x > 0$ it is true that

$$\mu \left\{ \frac{r\ell_{at}(s,u)}{Ba^{(\delta+1-\gamma)/\delta}\lambda^{1/\delta}} \geq x \right\} \to \frac{\sigma_t}{\delta x^\delta}. \tag{13.47}$$

We start with identity

$$\mu \left\{ \frac{r\ell_{at}(s,u)}{Ba^{(\delta+1-\gamma)/\delta}\lambda^{1/\delta}} \geq x \right\} = \lambda \int \int \mathbb{P} \left\{ R \geq \frac{Ba^{(\delta+1-\gamma)/\delta}\lambda^{1/\delta}x}{\ell_{at}(s,u)} \right\} F_U(du)ds.$$

By using inequality $\ell_{at}(s,u) \leq a$ and the high intensity condition we have

$$\frac{a^{(\delta+1-\gamma)/\delta}\lambda^{1/\delta}}{\ell_{at}(s,u)} \geq \left(\frac{\lambda}{a^{\gamma-1}} \right)^{1/\delta} \to \infty$$

uniformly in s and u. Hence, the tail asymptotics (12.7) for R applies, and we obtain

$$\mu \left\{ \frac{r\ell_{at}(s,u)}{Ba^{(\delta+1-\gamma)/\delta}\lambda^{1/\delta}} \geq x \right\} \sim \lambda c_R \int \int \left(\frac{Ba^{(\delta+1-\gamma)/\delta}\lambda^{1/\delta}x}{\ell_{at}(s,u)} \right)^{-\delta} F_U(du)ds$$

$$= \frac{c_R}{B^\delta} \int \int \frac{\ell_{at}(s,u)^\delta}{a^{\delta+1-\gamma}} F_U(du)ds \cdot x^{-\delta}.$$

Next, perform a variable change $s = a\tilde{s}$, $u = a\tilde{u}$ and use the self-similarity formula

$$\ell_{at}(s, u) = a\,\ell_t(\tilde{s}, \tilde{u}). \tag{13.48}$$

We obtain

$$\int \int \frac{\ell_{at}(s, u)^\delta}{a^{\delta+1-\gamma}} F_U(du)\,ds = a^\gamma \int \int \ell_t(\tilde{s}, \tilde{u})^\delta F_{U/a}(d\tilde{u})\,d\tilde{s}.$$

Notice that by (12.6) the measures $a^\gamma F_{U/a}(d\tilde{u})$ converge weakly to the measure $\frac{c_U\,\gamma\,d\tilde{u}}{\tilde{u}^{1+\gamma}}$ outside of any neighborhood of zero, since for any $y > 0$

$$a^\gamma F_{U/a}[y, \infty) = a^\gamma \mathbb{P}\{U \geq y\,a\} \sim a^\gamma c_U (y\,a)^{-\gamma}$$

$$= c_U y^{-\gamma} = \int_y^\infty \frac{c_U\,\gamma\,d\tilde{u}}{\tilde{u}^{1+\gamma}}. \tag{13.49}$$

It follows (omitting some evaluations at zero) that for any fixed $\tilde{s}$ we have

$$a^\gamma \int \ell_t(\tilde{s}, \tilde{u})^\delta F_{U/a}(d\tilde{u}) \to c_U\,\gamma \int \ell_t(\tilde{s}, \tilde{u})^\delta \frac{d\tilde{u}}{\tilde{u}^{1+\gamma}}.$$

It remains to justify the integration over $\tilde{s}$. Indeed, we may apply Lebesgue dominated convergence theorem because there is an integrable majorant

$$a^\gamma \int \ell_t(\tilde{s}, \tilde{u})^\delta F_{U/a}(d\tilde{u}) \leq \begin{cases} 0, & \tilde{s} > 1, \\ a^\gamma \mathbb{E}\,\min((\tfrac{U}{a})^\delta, 1) \leq c_1, & -1 \leq \tilde{s} \leq 1, \\ a^\gamma \mathbb{P}(U/a \geq \tilde{s}) \leq c_2 |\tilde{s}|^{-\gamma}, & \tilde{s} < -1. \end{cases}$$

Integration yields

$$a^\gamma \int \int \ell_t(\tilde{s}, \tilde{u})^\delta F_{U/a}(d\tilde{u})\,d\tilde{s} \to c_U\,\gamma \int \int \ell_t(\tilde{s}, \tilde{u})^\delta \frac{d\tilde{u}}{\tilde{u}^{1+\gamma}}\,d\tilde{s},$$

and we obtain

$$\mu\left\{ \frac{r\ell_{at}(s, u)}{Ba^{(\delta+1-\gamma)/\delta}\lambda^{1/\delta}} \geq x \right\} \to \frac{c_R c_U\,\gamma}{B^\delta} \int \int \ell_t(\tilde{s}, \tilde{u})^\delta \frac{d\tilde{u}}{\tilde{u}^{1+\gamma}}\,d\tilde{s} \cdot x^{-\delta}$$

$$= \frac{c_R c_U\,\gamma\sigma_t}{B^\delta x^\delta}.$$

The definition of B permits to simplify the constant and to obtain (13.47), as required.

The uniform bound (8.8) can be established by the same calculation. $\square$

Remark 13.15. The role of the assumption (13.19) requires some explanation. Let us fix the "reduced" parameters $\tilde{s}, \tilde{u}$. The corresponding expression $\ell_{at}(a\tilde{s}, a\tilde{u})$, has growth order a. Thus a barrier to which the variable R is compared, has the order $\frac{a^{(\delta+1-\gamma)/\delta}\lambda^{1/\delta}}{a} = \left(a^{-(\gamma-1)}\lambda\right)^{1/\delta}$. The latter expression tends to infinity exactly when (13.19) holds.

13.5.2 Convergence to a Poisson Telecom process

We will prove a limit theorem for the workload Z_a in the case of *critical intensity*

$$\frac{\lambda}{a^{\gamma-1}} \to L, \qquad 0 < L < \infty. \tag{13.50}$$

Theorem 13.16. *Assume that $1 < \gamma < \delta \le 2$, $a \to \infty$, and that critical intensity condition (13.50) holds. Then with scaling $b := a$ it is true that*

$$Z_a \xrightarrow{f.d.d.} Y_{\gamma,R} \ ,$$

where the process $Y_{\gamma,R}(t)$ (called Poisson Telecom process) admits an integral representation

$$Y_{\gamma,R}(t) = \int_{\mathcal{R}} r\, \ell_t(s,u) \widetilde{N}'(ds, du, dr).$$

Here $\ell_t(s,u)$ is the kernel defined in (12.4) and $\widetilde{N}'$ is a centered Poisson random measure of intensity

$$\mu'(ds, du, dr) := \frac{L\, c_U\, \gamma\, ds\, du}{u^{1+\gamma}}\, F_R(dr).$$

Proof. We prove the convergence of one-dimensional distributions. Recall that

$$Z_a(t) = \int_{\mathcal{R}} \frac{r\ell_{at}(s,u)}{a}\, d\widetilde{N}.$$

We need to check that for any $x > 0$ it is true that

$$\mu\left\{\frac{r\ell_{at}(s,u)}{a} \ge x\right\} \to \mu'\left\{r\ell_t(s,u) \ge x\right\}. \tag{13.51}$$

Indeed, by applying the same variable change as in the previous subsection, $s = a\tilde{s}$, $u = a\tilde{u}$, and using the self-similarity formula (13.48), we obtain

$$\mu\left\{\frac{r\ell_{at}(s,u)}{a} \ge x\right\} = \lambda \int\int\int \mathbf{1}_{\{r\,\ell_{at}(s,u)\ge ax\}}\, F_U(du)\, ds\, F_R(dr)$$

$$= a\lambda \int\int\int \mathbf{1}_{\{r\,\ell_t(\tilde{s},\tilde{u})\ge x\}} F_{U/a}(d\tilde{u})\, d\tilde{s}\, F_R(dr)$$

$$\sim L \int\int\int \mathbf{1}_{\{r\,\ell_t(\tilde{s},\tilde{u})\ge x\}} a^\gamma F_{U/a}(d\tilde{u})\, d\tilde{s}\, F_R(dr),$$

where we used (13.50) at the last step.

As we showed in (13.49), the measures $a^\gamma F_{U/a}(d\tilde{u})$ converge weakly to the measure $\frac{c_U \gamma d\tilde{u}}{\tilde{u}^{1+\gamma}}$ outside of any neighborhood of zero. It follows (omitting some evaluations at zero) that for any fixed $\tilde{s}$ we have

$$\int 1_{\{r\,\ell_t(\tilde{s},\tilde{u})\geq x\}} a^\gamma F_{U/a}(d\tilde{u}) \to \int 1_{\{r\,\ell_t(\tilde{s},\tilde{u})\geq x\}} \frac{c_U \gamma d\tilde{u}}{\tilde{u}^{1+\gamma}}.$$

By integrating this limit over $\tilde{s}$ and r, we obtain

$$\mu\left\{\frac{r\ell_{at}(s,u)}{a} \geq x\right\} \to L \int \int \int 1_{\{r\,\ell_t(\tilde{s},\tilde{u})\geq x\}} \frac{c_U \gamma d\tilde{u}}{\tilde{u}^{1+\gamma}} d\tilde{s}\, F_R(dr)$$

$$= \mu'\left\{r\ell_t(\tilde{s},\tilde{u}) \geq x\right\},$$

as required in (13.51).

The convergence of multivariate distributions is proved along exactly the same lines. By Cramér–Wold Theorem 4.8 it is sufficient to prove that every univariate projection converges, i.e. for any real $c_1,\ldots,c_n$ and any $t_1,\ldots,t_n \in [0,1]$ we have

$$\sum_{j=1}^{n} c_j Z_a(t_j) = \int_{\mathcal{R}} \frac{r\sum_{j=1}^{n} c_j \ell_{at_j}(s,u)}{a}\, d\tilde{N} \Rightarrow \sum_{j=1}^{n} c_j Y_{\gamma,R}(t_j).$$

For doing this, just replace $\ell_{at_j}(s,u)$ with $\sum_{j=1}^{n} c_j \ell_{at_j}(s,u)$ in the proof of univariate convergence.   □

Remark 13.17. The diagram of limit theorems for the critical intensity case (13.50) looks like that of low intensity exposed on Figure 13.2, with the single change: in zone E γ-stable Lévy process is replaced by Telecom Poisson process from Theorem 13.16. Indeed, for cases $\gamma = \delta = 2$ (zone A, convergence to Wiener process) and $1 < \delta < \gamma \leq 2$ (zone D, convergence to a δ-stable Lévy process) Theorems 13.1 and 13.6 still apply.

Exercise 13.18. Prove that Poisson Telecom process $Y_{\gamma,R}$ is defined correctly, which means, according to (7.20), that the kernel

$$f_t(s,u,r) := r\,\ell_t(s,u)$$

must satisfy

$$\int_{\mathcal{R}} \min\{f_t, f_t^2\} \frac{ds\,du}{u^{1+\gamma}}\, F_R(dr) < \infty.$$

Exercise 13.19. Prove that both Telecom processes $\mathcal{Z}_{\gamma,\delta}$ and $Y_{\gamma,R}$ are self-similar.

We refer to [22, 35] for further studies of Poisson Telecom processes.

13.6 *Handling "messengers from the past"*

When one wishes to arrange computer modelling of the service systems with
long range dependence, the question of handling events relating "present
time" and "infinite past" turns out to be delicate. On one hand, full modelling of the long past intervals is inefficient because most of the events in
the distant past do not have influence on the present. On the other hand, if
we do not take the distant past into account at all we loose some "messages
from the past" that may be really important for the present.

In modelling of our simple system, we are lucky to have a nice solution
to this problem [49]. Namely, we may replace the integral workload coming
from service processes started before zero time with the workload coming
from another, even simpler model where all processes start at zero time.

This simpler model can be described as follows: let $\mathcal{R}_1 := \mathbb{R}_+ \times \mathbb{R}_+$ and
let μ_1 be a measure on $\mathcal{R}_1$ defined by

$$\mu_1(du, dr) = \lambda f(u) du P_R(dr)$$

with the density $f(u) := \mathbb{P}(U \geq u)$. Notice that by (1.6) the density f is
integrable iff $\mathbb{E}\,U < \infty$. Furthermore, let N_1 be a Poisson random measure
on $\mathcal{R}_1$ with intensity μ_1, and $\widetilde{N}_1$ the corresponding centered measure.

We interpret the elements $(u, r) \in \mathcal{R}_1$ as service processes starting at
time zero, lasting u units of time, and requiring r units of resources. We
stress that whenever $\mathbb{E}\,U < \infty$ we have $\mu_1(\mathcal{R}_1) < \infty$, hence the number of
these service processes $N_1(\mathcal{R}_1)$ is finite almost surely.

Let $V(t), t \geq 0$, be the part of integral workload (12.2) related to the
service processes already active at time zero, so to say "messengers from
the past",

$$V(t) := \int_{\mathcal{R}} r\ell_t(s, u)\, \mathbf{1}_{\{s \leq 0\}}\, d\widetilde{N}.$$

Proposition 13.20. *The processes $V(t)$ and*

$$V_1(t) := \int_{\mathcal{R}_1} r\ell_t(0, u)\, d\widetilde{N}_1, \qquad t \geq 0,$$

have the same finite-dimensional distributions.

Proof.   In the definition of V we clearly may restrict the integration do-
main to $\mathcal{Q} := \{(s, u, r) \in \mathcal{R} : s \leq 0, u > |s|\}$ since the integrand vanishes
on the complement of $\mathcal{Q}$.

Consider a mapping $J : \mathcal{Q} \mapsto \mathcal{R}_1$ defined by

$$J(s, u, r) = (u - |s|, r).$$

Since for any $t \geq 0, u \geq 0, s \leq 0$ we have

$$\ell_t(s, u) = \begin{cases} 0, & u < |s|, \\ \ell_t(0, u - |s|), & u \geq |s|, \end{cases}$$

we see that for each $t \geq 0$ the integrand in the definition of V taken at the point (s, u, r) is equal to the integrand in the definition of V_1 taken at the point $J(s, u, r)$. This allows us to write a variable change formula

$$V(t) = \int_{\mathcal{R}_1} r\ell_t(0, u) \, d\widetilde{N}_2, \qquad t \geq 0,$$

where $\widetilde{N}_2(A) := \widetilde{N}(J^{-1}(A))$ is a centered Poisson measure on $\mathcal{R}_1$ with intensity measure $\mu_2(A) := \mu(J^{-1}(A))$.

Now V and V_1 are written as Poisson integrals with equal kernels on the same space. It remains to show that the corresponding intensities coincide, i.e. $\mu_2 = \mu_1$. In other words, it is enough to prove that for any $x > 0$ it is true that

$$\mu\big((s, u, r) : s < 0, u > |s|, (r, u - |s|) \in B\big) = \mu_1\big((u, r) \in B\big), \qquad B \subset \mathcal{B}^2.$$

Since r is an independent coordinate with the same distribution in both measures, this boils down to

$$\mu\big((s, u, r) : s < 0, u > |s|, u - |s| \in B\big) = \mu_1\big((u, r) : u \in B\big), \qquad B \subset \mathcal{B}^1.$$

In turn, this can be done by comparing distribution functions, i.e. by showing

$$\mu\big((s, u, r) : s < 0, u - |s| > x\big) = \mu_1\big((u, r) : u > x\big)$$

for any $x > 0$. Indeed, we have

$$\mu\big((s, u, r) : s < 0, u - |s| > x\big) = \lambda \int_0^\infty \left(\int_0^\infty \mathbf{1}_{\{s < u - x\}} ds \right) F_U(du)$$

$$= \lambda \int_x^\infty (u - x) F_U(du)$$

and

$$\mu_1\big((u, r) : u > x\big) = \lambda \int_x^\infty f(v) dv$$

$$= \lambda \int_x^\infty \left(\int_x^\infty \mathbf{1}_{\{u \geq v\}} F_U(du) \right) dv$$

$$= \lambda \int_x^\infty \left(\int_x^\infty \mathbf{1}_{\{u \geq v\}} dv \right) F_U(du)$$

$$= \lambda \int_x^\infty (u - x) F_U(du),$$

exactly as above. $\qquad\qquad\qquad\qquad\qquad\qquad\qquad\qquad\qquad\qquad\square$

The conclusion is as follows. Instead of modelling process V related to the infinite horizon of variable s, we may perform a compact procedure based on Proposition 7.1: pick up a Poisson random variable M of intensity $\lambda \mathbb{E} U$ (the number of messengers from the past), then choose M random variables with $u_1, \ldots, u_M$ with common distribution density $\frac{f(\cdot)}{\mathbb{E} U}$, jointly with independent copies of resource variables $r_1, \ldots, r_M$, and include triplets $(0, u_i, r_i)$ in the model, i.e add service processes starting at time 0, lasting u_i units of time, and requiring r_i units of resources.

14 Micropulse Model

The *micropulse model* introduced by Cioczek-Georges and Mandelbrot [18, 19] can be described as follows. A system registers independently arriving *micropulses*. Every micropulse is characterized by its *size* $r \geq 0$, *arrival time* $s \in \mathbb{R}$, and *duration* $u \geq 0$. At instant $s + u$ micropulse is cancelled, in other words, a micropulse of size $-r$ arrives in the system. The problem is to evaluate the sum of non-cancelled micropulses on the long time intervals.

The formal model is based on Poisson random measure. Let
$$\mathcal{R} := \{(s, u, r)\} = \mathbb{R} \times \mathbb{R}_+ \times \mathbb{R}_+.$$
Every point $(s, u, r) \in \mathcal{R}$ corresponds to an arriving micropulse.

The system is characterized by the following parameters:

- $\lambda > 0$ – arrival intensity of micropulses, i.e. the average number of micropulses arriving during any time interval of unit length;
- $F_U(du)$ – the distribution of micropulse duration;
- $F_R(dr)$ – the distribution of micropulse size.

Define on $\mathcal{R}$ intensity measure
$$\mu(ds, du, dr) := \lambda \, ds \, F_U(du) \, F_R(dr).$$

Let N be the corresponding Poisson random measure. One may consider the samples of N (sets of triplets (s, u, r), each triplet corresponding to a micropulse) as variants (sample paths) of system's work. We are able to express many system characteristics as Poisson integrals with respect to N.

In particular, we are interested in the workload, this time understood as the sum of micropulses on the family of intervals $[-t, t]$:

$$W^{(m)}(t) := \int_{\mathcal{R}} r \left(\mathbf{1}_{\{ -t \leq s \leq t \atop t < s+u \}} - \mathbf{1}_{\{ s < -t \atop -t \leq s+u \leq t \}} \right) dN. \tag{14.1}$$

Here the first indicator corresponds to the micropulses that arrived during the interval $[-t, t]$ and were not cancelled before time t, while the second one corresponds to the micropulses that arrived before time $-t$ and were cancelled during $[-t, t]$.

We study the behavior of $W^{(m)}(\cdot)$ on large time intervals $[0, a]$, $a \to \infty$. Scaling by appropriate factor $b = b(a, \lambda)$ leads to a normalized workload process,

$$Z_a^{(m)}(t) := \frac{W^{(m)}(at)}{b}, \qquad t \in [0, 1]. \tag{14.2}$$

Here a and λ are considered as variables (at least one of them should tend to infinity), and the normalizing parameter b depends on a, λ, and of distributions F_U and F_R.

Cioczek-Georges and Mandelbrot proved that in certain regimes $Z_a^{(m)}$ behaves like fractional Brownian motion. Even more interestingly, Marouby [69] showed that $Z_a^{(m)}$ may asymptotically behave like *bifractional Brownian motion* $B^{H,K}$ with $H = \frac{1}{2}$. Recall that $B_t^{H,K}, t \geq 0$, is a centered Gaussian process with covariance

$$\mathbb{E}\left(B_t^{H,K} B_s^{H,K} \right) = 2^{-K} \left((t^{2H} + s^{2H})^K - |t - s|^{2HK} \right) \tag{14.3}$$

with the set of admissible parameters

$$\{H, K : 0 < H \leq 1, 0 < K \leq 2, HK \leq 1\}.$$

Theorem 14.1. *Let* $\mathbb{E} R^2 < \infty$. *Assume that for some* $\gamma \in (0, 1)$ *it is true that*

$$\mathbb{P}(U > u) \sim c_U u^{-\gamma}, \qquad u \to \infty, \tag{14.4}$$

$a \to \infty$, *and high intensity condition holds*

$$\lambda a^{1-\gamma} \to \infty. \tag{14.5}$$

Then the scaling

$$b = b(a, \lambda) := \left(\frac{\gamma \, 2^{2-\gamma} \, c_U \, \mathbb{E} R^2}{1 - \gamma} \, a^{1-\gamma} \lambda \right)^{1/2}$$

yields

$$Z_a^{(m)} \xrightarrow{f.d.d.} B^{1/2, 1-\gamma}. \tag{14.6}$$

Proof. We first investigate expectation and covariance of the workload $Z_a^{(m)}(\cdot)$ and show how formula (14.3) with appropriate H and K appears.

For expectation we have

$$\mathbb{E}\, Z_a^{(m)}(t) = \frac{1}{b}\, \mathbb{E}\int_{\mathcal{R}} r\left(\mathbf{1}_{\{\substack{-at\leq s\leq at\\ at<s+u}\}} - \mathbf{1}_{\{\substack{s<-at\\ -at\leq s+u\leq at}\}}\right)dN$$

$$= \frac{\lambda}{b}\int r\int\int\left(\mathbf{1}_{\{\substack{-at\leq s\leq at\\ at<s+u}\}} - \mathbf{1}_{\{\substack{s<-at\\ -at\leq s+u\leq at}\}}\right)ds F_U(du)F_R(dr).$$

Notice that

$$\int\mathbf{1}_{\{\substack{-at\leq s\leq at\\ at<s+u}\}}\,ds = \begin{cases} 2at, & \text{if } u\geq 2at,\\ u, & \text{if } u<2at; \end{cases}$$

$$\int\mathbf{1}_{\{\substack{s<-at\\ -at\leq s+u\leq at}\}}\,ds = \begin{cases} 2at, & \text{if } u\geq 2at,\\ u, & \text{if } u<2at. \end{cases}$$

Therefore, for any $u\geq 0$ the interior integral vanishes and we obtain

$$\mathbb{E}\, Z_a^{(m)}(t) = 0.$$

Now we find the covariance of $Z_a^{(m)}$. Let $t_1\leq t_2$. We have

$$\mathrm{cov}(Z_a^{(m)}(t_1), Z_a^{(m)}(t_2)) = \mathbb{E}\left(Z_a^{(m)}(t_1)Z_a^{(m)}(t_2)\right)$$

$$= \frac{\lambda}{b^2}\int_{\mathcal{R}} r^2\int\int A(s,u)\, B(s,u)ds F_U(du)F_R(dr),$$

where

$$A(s,u) := \mathbf{1}_{\{\substack{-at_1\leq s\leq at_1\\ at_1<s+u}\}} - \mathbf{1}_{\{\substack{s<-at_1\\ -at_1\leq s+u\leq at_1}\}},$$

$$B(s,u) := \mathbf{1}_{\{\substack{-at_2\leq s\leq at_2\\ at_2<s+u}\}} - \mathbf{1}_{\{\substack{s<-at_2\\ -at_2\leq s+u\leq at_2}\}}.$$

Notice that (see Figure 14.1)

$$A(s,u) = B(s,u) = \begin{cases} 1, & \text{if } -at_1\leq s\leq at_1\leq at_2<s+u;\\ -1, & \text{if } s<-at_2\leq -at_1\leq s+u\leq at_1. \end{cases}$$

Otherwise, either $A(s,u) = 0$, or $B(s,u) = 0$.

For any fixed $u\geq 0$ the set $S_1(u) := \{s\,:\, A(s,u) = B(s,u) = 1\}$ represents an interval

$$S_1(u) = \begin{cases} \emptyset, & \text{if } 0\leq u\leq at_2 - at_1;\\ (at_2 - u, at_1], & \text{if } at_2 - at_1 < u\leq at_2 + at_1;\\ [-at_1, at_1], & \text{if } u > at_2 + at_1. \end{cases}$$

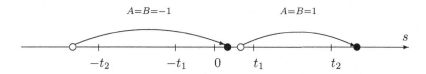

Fig. 14.1   Micropulses important for variance calculation.

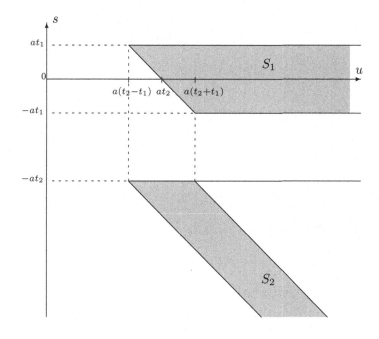

Fig. 14.2   Sets S_1 and S_2.

Similarly, for $S_2(u) := \{s : A(s,u) = B(s,u) = -1\}$ we find (see Figure 14.2)

$$S_2(u) = \begin{cases} \emptyset, & \text{if } 0 \le u \le at_2 - at_1; \\ [-at_1 - u, -at_2), & \text{if } at_2 - at_1 < u \le at_2 + at_1; \\ [-at_1 - u, at_1 - u], & \text{if } u > at_2 + at_1. \end{cases}$$

Therefore, for each $u \geq 0$ we have

$$\int A(s,u)B(s,u)ds = |S_1(u) \cup S_2(u)|$$

$$= \begin{cases} 0, & \text{if } 0 \leq u \leq at_2 - at_1; \\ 2(at_1 - at_2 + u), & \text{if } at_2 - at_1 \leq u \leq at_2 + at_1; \\ 4at_1, & \text{if } u \geq at_2 + at_1. \end{cases}$$

It follows that

$$\mathrm{cov}(Z_a^{(m)}(t_1), Z_a^{(m)}(t_2))$$

$$= \frac{\lambda}{b^2} \, \mathbb{E}R^2 \left(2 \int_{a(t_2-t_1)}^{a(t_2+t_1)} (a(t_1 - t_2) + u)F_U(du) + 4at_1 \, \mathbb{P}(U > a(t_2 + t_1)) \right)$$

$$= \frac{\lambda}{b^2} \, \mathbb{E}R^2 \left(2 \int_{a(t_2-t_1)}^{a(t_2+t_1)} uF_U(du) + 2a(t_1 + t_2) \, \mathbb{P}(U > a(t_2 + t_1)) \right.$$

$$\left. -2a(t_2 - t_1)\mathbb{P}(U > a(t_2 - t_1)) \right).$$

Integrating by parts and assumption (14.4) yield

$$\int_{a(t_2-t_1)}^{a(t_2+t_1)} uF_U(du) = u \, \mathbb{P}(U > u) \Big|_{u=a(t_2+t_1)}^{u=a(t_2-t_1)} + \int_{a(t_2-t_1)}^{a(t_2+t_1)} \mathbb{P}(U > u) \, du$$

$$= a(t_2 - t_1)\mathbb{P}(U > a(t_2 - t_1)) - a(t_2 + t_1)\mathbb{P}(U > a(t_2 + t_1))$$

$$+ c_U \int_{a(t_2-t_1)}^{a(t_2+t_1)} u^{-\gamma} du (1 + o(1))$$

$$= a(t_2 - t_1)\mathbb{P}(U > a(t_2 - t_1)) - a(t_2 + t_1)\mathbb{P}(U > a(t_2 + t_1))$$

$$+ \frac{c_U a^{1-\gamma}}{1-\gamma} \left((t_2 + t_1)^{1-\gamma} - (t_2 - t_1)^{1-\gamma} \right) (1 + o(1)).$$

Plugging into the previous estimate leads to cancellation of the boundary terms. By the definition of b, we obtain

$$\mathrm{cov}(Z_a^{(m)}(t_1), Z_a^{(m)}(t_2)) \to \frac{2\lambda}{b^2} \, \mathbb{E}R^2 \, \frac{c_U a^{1-\gamma}}{1-\gamma} \left((t_2 + t_1)^{1-\gamma} - (t_2 - t_1)^{1-\gamma} \right)$$

$$= 2^{-(1-\gamma)} \left((t_2 + t_1)^{1-\gamma} - (t_2 - t_1)^{1-\gamma} \right) = \mathrm{cov}\left(B_{t_1}^{1/2, 1-\gamma}, B_{t_2}^{1/2, 1-\gamma} \right). \quad (14.7)$$

Of course, this is in agreement with our theorem's assertion.

The remainder of the proof repeats the arguments used in the proof of Theorem 13.4. We reproduce them for completeness. In order to verify (14.6), by Cramér–Wold Theorem 4.8 it is sufficient to prove that every univariate projection converges, i.e. for any real $c_1, \ldots, c_n$ and any $t_1, \ldots, t_n \in [0, 1]$ we have

$$\sum_{j=1}^n c_j Z_a^{(m)}(t_j) \Rightarrow \sum_{j=1}^n c_j B^{1/2, 1-\gamma}(t_j). \tag{14.8}$$

Recall that by (14.2) and (14.1) for each $j \leq n$ we have

$$Z_a^{(m)}(t_j) = \int_{\mathcal{R}} f_{t_j, a} dN,$$

where

$$f_{t,a}(s, u, r) := \frac{r}{b} \left(1_{\{ {-t \leq s \leq t \atop t < s + u} \}} - 1_{\{ {s < -t \atop -t \leq s + u \leq t} \}} \right) := \frac{r}{b} \Delta_{t,a}(s, u).$$

Then (14.8) writes as

$$S_a := \int_{\mathcal{R}} \left(\sum_{j=1}^n c_j f_{t_j, a} \right) dN \Rightarrow \sum_{j=1}^n c_j B^{1/2, 1-\gamma}(t_j) := S.$$

The proof of convergence of centered Poisson integrals to a normal distribution is based on Proposition 8.1. First, we must check that

$$\mathbb{V}ar S_a \to \mathbb{V}ar S;$$

Indeed, by (14.7)

$$\mathbb{V}ar S_a = \mathbb{V}ar \sum_{j=1}^n c_j Z_a^{(m)}(t_j)$$

$$= \sum_{i,j=1}^n c_i c_j \, \text{cov}(Z_a^{(m)}(t_i), Z_a^{(m)}(t_j))$$

$$\to \sum_{i,j=1}^n c_i c_j \, \text{cov}(B^{1/2, 1-\gamma}(t_i), B^{1/2, 1-\gamma}(t_j))$$

$$= \mathbb{V}ar \sum_{j=1}^n c_j B^{1/2, 1-\gamma}(t_j) = \mathbb{V}ar S.$$

Now it remains to verify the second assumption (8.3) of Proposition 8.1, i.e. we check that for any $\varepsilon > 0$

$$\int_{\{|f_a| > \varepsilon\}} f_a^2 \, d\mu \to 0, \tag{14.9}$$

where

$$f_a := \sum_{j=1}^{n} c_j f_{t_j,a} = \sum_{j=1}^{n} c_j \frac{r}{b} \Delta_{t_j,a}(s,u).$$

Clearly, we have

$$0 \le |f_a| \le \max_{1 \le j \le n} |c_j| \frac{r}{b} \sum_{j=1}^{n} |\Delta_{t_j,a}(s,u)| := C \frac{r}{b} \sum_{j=1}^{n} |\Delta_{t_j,a}(s,u)|,$$

hence

$$\{|f_a| > \varepsilon\} \subset \left\{ r > \frac{\varepsilon b}{Cn} \right\}$$

and

$$|f_a|^2 \le C^2 \frac{r^2}{b^2} n \sum_{j=1}^{n} \Delta_{t_j,a}(s,u)^2.$$

It follows that

$$\int_{\{|f_a|>\varepsilon\}} f_a^2 \, d\mu \le C^2 n \sum_{j=1}^{n} \int \mathbf{1}_{\{r>\frac{\varepsilon b}{Cn}\}} \frac{r^2}{b^2} \Delta_{t_j,a}(s,u)^2 \, d\mu.$$

$$= \frac{C^2 n}{b^2} \mathbb{E}\left(R^2 \mathbf{1}_{\{R>\frac{\varepsilon b}{Cn}\}}\right) \lambda \sum_{j=1}^{n} \int \int \Delta_{t_j,a}(s,u)^2 \, ds F_U(du)$$

$$= \frac{C^2 n \mathbb{E}\left(R^2 \mathbf{1}_{\{R>\frac{\varepsilon b}{Cn}\}}\right)}{\mathbb{E}R^2} \sum_{j=1}^{n} \mathbb{E}\, Z_a^{(m)}(t_j)^2.$$

By (14.5) b tends to infinity, thus $\mathbb{E}\left(R^2 \mathbf{1}_{\{R>\frac{\varepsilon b}{Cn}\}}\right)$ tends to zero. By (14.7) we have

$$\mathbb{E}\, Z_a^{(m)}(t_j)^2 \to \mathbb{E}\, B^{1/2,1-\gamma}(t_j)^2 = t_j^{1-\gamma} < \infty.$$

Hence, (14.9) follows and we are done.   □

Our next goal is to find a stable analogue to bifractional Brownian motion. One may define such analogue as a workload scaling limit in a setting similar to Theorem 13.13 which had led us to a stable Telecom process. This setting would differ from that of Theorem 14.1 by the distribution of micropulse size R. Namely, in addition to (14.4) one should assume that

$$\mathbb{P}(R > r) \sim c_R r^{-\delta}, \qquad r \to \infty,$$

for some $\delta \in (\gamma, 2)$ and $c_R > 0$.

Let us choose a scaling factor as

$$b = b(a, \lambda) := B \left(a^{1-\gamma}\lambda\right)^{1/\delta} \qquad (14.10)$$

where the constant B is defined by $B^\delta := c_R c_U \delta \gamma$.

Theorem 14.2. *Let $0 < \gamma < 1$, $\gamma < \delta < 2$. Assume that $a \to \infty$ and that high intensity condition (14.5) holds. Then with the scaling (14.10) it is true that*

$$Z_a^{(m)} \xrightarrow{\text{f.d.d.}} \mathcal{Z}_{\gamma,\delta}^{(m)},$$

and the δ-stable process $\mathcal{Z}_{\gamma,\delta}^{(m)}(t)$ admits an integral representation

$$\mathcal{Z}_{\gamma,\delta}^{(m)}(t) = \int_{\mathcal{R}} r \left(\mathbf{1}_{\{\substack{-t \le s \le t \\ t < s + u}\}} - \mathbf{1}_{\{\substack{s < -t \\ -t \le s + u \le t}\}}\right) dN', \qquad (14.11)$$

where N' is a Poisson random measure of intensity

$$\mu' := u^{-\gamma-1} r^{-\delta-1} \, ds \, du \, dr.$$

It is natural to call $\mathcal{Z}_{\gamma,\delta}^{(m)}(\cdot)$ *bifractional stable process.*

Proof. We only give a sketch. Recall that by (14.1) and (14.2)

$$Z_a^{(m)}(t) = \int_{\mathcal{R}} \frac{r}{b} \left(\mathbf{1}_{\{\substack{-at \le s \le at \\ at < s + u}\}} - \mathbf{1}_{\{\substack{s < -at \\ -at \le s + u \le at}\}}\right) dN.$$

Making a variable change $u = a\tilde{u}$, $s = a\tilde{s}$, $r = b\tilde{r}$, we have

$$Z_a^{(m)}(t) = \int_{\mathcal{R}} \tilde{r} \left(\mathbf{1}_{\{\substack{-t \le \tilde{s} \le t \\ t < \tilde{s} + \tilde{u}}\}} - \mathbf{1}_{\{\substack{\tilde{s} < -t \\ -t \le \tilde{s} + \tilde{u} \le t}\}}\right) dN_a',$$

where N_a' is a Poisson random measure of intensity

$$
\begin{aligned}
\mu_a' &= a\,\lambda\,d\tilde{s}\,F_{U/a}(d\tilde{u})\,F_{R/b}(d\tilde{r}) \\
&= d\tilde{s}\,\left(\lambda^{1-\gamma}ab^{-\delta}\right)\lambda^\gamma F_{U/a}(d\tilde{u})\,b^\delta F_{R/b}(d\tilde{r}) \\
&= d\tilde{s}\,B^{-\delta}\,\lambda^\gamma F_{U/a}(d\tilde{u})\,b^\delta F_{R/b}(d\tilde{r}).
\end{aligned}
$$

Recall that measures $a^\gamma F_{U/a}(d\tilde{u})$ converge weakly to the measure $\frac{c_U \gamma \, d\tilde{u}}{\tilde{u}^{1+\gamma}}$ outside any neighborhood of zero, whenever $a \to \infty$. Similarly, since by (14.5) $b \to \infty$, the measures $b^\delta F_{R/b}(d\tilde{r})$ converge weakly to the measure $\frac{c_R \delta \, d\tilde{r}}{\tilde{r}^{1+\delta}}$ outside any neighborhood of zero.

Therefore, the intensities μ_a' converge to a μ' in a reasonable sense and convergence of processes follows. By Exercise 7.3 the limiting process is δ-stable. $\qquad\square$

Remark 14.3. For any $t \ge 0$ the distribution of $Z_{\gamma,\delta}^{(m)}(t)$ is *symmetric stable*, because positive and negative indicators in (14.11) are well balanced. It does not mean, however, that all linear combinations of these variables have symmetric distributions.

15   Spacial Extensions

15.1   *Spacial model*

In many problems such as wireless network modelling or modelling of porous media, the *spacial* aspect of observed phenomena is very important. This is why the traffic models we have considered so far, merit a spacial extension. Formally, we merely replace the temporal parameter $s \in \mathbb{R}$ with the spacial parameter $\mathbf{s} \in \mathbb{R}^d$. Accordingly, in the sequel $d\mathbf{s}$ means Lebesgue integration in $\mathbb{R}^d$ and $|A|$ stands for Lebesgue measure of a set $A \subset \mathbb{R}^d$.

The formal model of the system based on Poisson random measures looks as follows. Let $\mathcal{R} := \{(\mathbf{s}, u, r)\} = \mathbb{R}^d \times \mathbb{R}_+ \times \mathbb{R}_+$. Every point $(\mathbf{s}, u, r)$ corresponds to a *grain* with *anchor point* $\mathbf{s}$, *size* u, and *strength* r.

For example, one may interpret $\mathbf{s}$ as a location of a node of wireless network, u as the signal emission range, and r as the signal strength.

The system is characterized by the following parameters:

- $\lambda > 0$ – arrival intensity of anchor points, i.e. the average number of grains anchored at any spacial set of unit volume;
- $F_U(du)$ – the distribution of grain size;
- $F_R(dr)$ – the distribution of grain strength;
- G – grain shape, a bounded set of unit volume in $\mathbb{R}^d$.

Define on $\mathcal{R}$ an intensity measure

$$\mu(d\mathbf{s}, du, dr) = \lambda \, d\mathbf{s} \, F_U(du) \, F_R(dr).$$

Let N be a corresponding Poisson random measure and $\widetilde{N}$ its centered version. One may consider the samples of random measure N (sets of triplets $(\mathbf{s}, u, r)$, each triplet corresponding to a grain G scaled by u and shifted to the anchor point $\mathbf{s}$) as variants (sample paths) of the system configuration.

Random grain models (often with the constant strength parameter) were considered in [8–10, 13, 32, 48]. We reproduce here, in a modified form, some results from [48].

The local load on the system at a point $\mathbf{t} \in \mathbb{R}^d$ is understood as the total strength of all grains covering $\mathbf{t}$ and writes as

$$W^\circ(\mathbf{t}) = \int_{\mathcal{R}} r \mathbf{1}_{\{\mathbf{t} \in \mathbf{s} + uG\}} \, dN, \qquad (15.1)$$

while the integral load over a set $A \subset \mathbb{R}^d$ is

$$W^*(A) := \int_A W^\circ(\mathbf{t})dt = \int_\mathcal{R} r \int_A \mathbf{1}_{\{\mathbf{t} \in \mathbf{s} + uG\}} dt\, dN$$

$$= \int_\mathcal{R} r \cdot \left| A \cap (\mathbf{s} + uG) \right| dN. \tag{15.2}$$

Sometimes it is more convenient to measure the integral load not on a set but with respect to a function $\varphi : \mathbb{R}^d \mapsto \mathbb{R}$,

$$W^*(\varphi) := \int W^\circ(\mathbf{t})\varphi(\mathbf{t})dt = \int_\mathcal{R} r \int \varphi(\mathbf{t}) \mathbf{1}_{\{\mathbf{t} \in \mathbf{s} + uG\}} dt\, dN. \tag{15.3}$$

Notice that $W^*(\varphi)$ is linear in argument φ. The former expression (15.2) reappears if we let $\varphi = \mathbf{1}_{\{A\}}$.

The advantage of (15.3) is especially appreciated in cases when the local load integral (15.1) is not well defined but the integral (15.3) is well defined for sufficiently smooth functions φ.

For expectation, assuming that φ is integrable, that $\mathbb{E}\left(U^d\right) < \infty$ and $\mathbb{E}R < \infty$, using $|G| = 1$, we have

$$\mathbb{E}\,W^*(\varphi) = \int_\mathcal{R} r \int \varphi(\mathbf{t}) \mathbf{1}_{\{\mathbf{t} \in \mathbf{s} + uG\}} dt\, \mu(d\mathbf{s},\, du,\, dr)$$

$$= \lambda \int rF_R(dr) \int \varphi(\mathbf{t}) \int \int \mathbf{1}_{\{\mathbf{s} \in \mathbf{t} - uG\}} d\mathbf{s}\, F_U(du)\, dt$$

$$= \lambda\, \mathbb{E}R \int \varphi(\mathbf{t}) \int u^d |G| F_U(du)\, dt$$

$$= \lambda \cdot \mathbb{E}R \cdot \mathbb{E}\left(U^d\right) \cdot \int_{\mathbb{R}^d} \varphi(\mathbf{t})dt.$$

Similarly, under natural assumptions $\varphi_1, \varphi_2 \in \mathbb{L}_1(\mathbb{R}^d)$, $\mathbb{E}\left(U^d\right) < \infty$ and $\mathbb{E}R^2 < \infty$ we have for covariance

$$\mathrm{cov}(W^*(\varphi_1), W^*(\varphi_2))$$

$$= \int_\mathcal{R} r^2 \left(\int \varphi_1(\mathbf{t}_1) \mathbf{1}_{\{\mathbf{t}_1 \in \mathbf{s} + uG\}} dt_1 \int \varphi_2(\mathbf{t}_2) \mathbf{1}_{\{\mathbf{t}_2 \in \mathbf{s} + uG\}} dt_2 \right) d\mu \tag{15.4}$$

$$= \lambda\, \mathbb{E}R^2 \int \int \varphi_1(\mathbf{t}_1)\varphi_2(\mathbf{t}_2) \int |uG \cap ((t_2 - t_1) + uG)|\, F_U(du)\, dt_1 dt_2.$$

In particular,

$$\mathbb{V}ar(W^*(\varphi))$$

$$= \lambda\, \mathbb{E}\,R^2 \int \int \varphi(\mathbf{t}_1)\varphi(\mathbf{t}_2) \int |uG \cap ((t_2 - t_1) + uG)|\, F_U(du)\, dt_1 dt_2.$$

Remark 15.1. One may obtain more flexible model by replacing dichotomy $\mathbf{t} \in \mathbf{s} + uG$ *vs.* $\mathbf{t} \notin \mathbf{s} + uG$ in the load formula (15.1), that is characteristic for grain models, by a functional dependence, see e.g. [14, 46]. Let $h : \mathbb{R}^d \mapsto \mathbb{R}_+$ be a *fading function* going to zero when the argument goes to infinity. Roughly speaking, this function shows how the strength of the signal depends on the relative location of the emitting node and reception point. Then we may extend (15.1) and (15.3) to

$$W_h^\circ(\mathbf{t}) := \int_{\mathcal{R}} rh\left(\frac{\mathbf{t} - \mathbf{s}}{u}\right) dN,$$

$$W_h^*(\varphi) := \int_{\mathcal{R}} r \int \varphi(\mathbf{t}) h\left(\frac{\mathbf{t} - \mathbf{s}}{u}\right) dt\, dN.$$

15.2 *Spacial noise integrals*

Before passing to limit theorems for spacial load processes let us introduce some Gaussian processes that will show up as limiting objects later on.

15.2.1 *White noise integral*

A centered Gaussian process

$$\mathbb{W}(\varphi), \qquad \varphi \in \mathbb{L}_2(\mathbb{R}^d),$$

with covariance

$$\mathbb{E}\,\mathbb{W}(\varphi)\mathbb{W}(\psi) = \int_{\mathbb{R}^d} \varphi(\mathbf{t})\psi(\mathbf{t}) dt \tag{15.5}$$

is called *white noise integral.* We stress that according to (7.2) this process can be obtained as a white noise integral with respect to a white noise $\mathcal{W}$ on $\mathbb{R}^d$ with Lebesgue intensity,

$$\mathbb{W}(\varphi) = \int_{\mathbb{R}^d} \varphi(\mathbf{t}) d\mathcal{W}.$$

There is a simple connection of the white noise integral to Wiener process and to Brownian sheet. Recall that the process $W(t) := \mathbb{W}(\mathbf{1}_{\{[0,t]\}})$ is a Brownian sheet (cf. Subsection 7.3); for $d = 1$ we obtain a Wiener process.

15.2.2 *Fractional noise integral*

Before we define a fractional noise integral, we need to describe a class of appropriately integrable functions.

For a function $\varphi \in \mathbb{L}_1(\mathbb{R}^d)$ define its *Riesz energy* of order $\alpha \in (0,1)$ by

$$E_\alpha(\varphi) := \int_{\mathbb{R}^d} \int_{\mathbb{R}^d} \frac{\varphi(\mathbf{t}_1)\varphi(\mathbf{t}_2)}{|\mathbf{t}_1 - \mathbf{t}_2|^{(1-\alpha)d}} \, d\mathbf{t}_1 d\mathbf{t}_2 \ \in [0, +\infty). \tag{15.6}$$

Let

$$\mathbb{M}_\alpha := \left\{ \varphi \in \mathbb{L}_1(\mathbb{R}^d) : E_\alpha(\varphi) < +\infty \right\}$$

be the space of functions of finite Riesz energy. We may supply $\mathbb{M}_\alpha$ with a natural scalar product

$$[\varphi, \psi]_\alpha := \int_{\mathbb{R}^d} \int_{\mathbb{R}^d} \frac{\varphi(\mathbf{t}_1)\psi(\mathbf{t}_2)}{|\mathbf{t}_1 - \mathbf{t}_2|^{(1-\alpha)d}} d\mathbf{t}_1 d\mathbf{t}_2.$$

This scalar product admits a convenient representation in terms of *Fourier transform*, cf. [59], Chapter 5. Namely, if $\varphi \in \mathbb{M}_\alpha$, then its Fourier transform

$$\widehat{\varphi}(\mathbf{u}) := \int_{\mathbb{R}^d} e^{-2\pi i(\mathbf{u},\mathbf{t})} \varphi(\mathbf{t}) d\mathbf{t}$$

satisfies

$$\int_{\mathbb{R}^d} |\widehat{\varphi}(\mathbf{u})|^2 \, |\mathbf{u}|^{-\alpha d} d\mathbf{u} < +\infty$$

and for any $\varphi, \psi \in \mathbb{M}_\alpha$ we have

$$[\varphi, \psi]_\alpha = c_{\alpha,d} \int_{\mathbb{R}^d} \widehat{\varphi}(\mathbf{u}) \, \overline{\widehat{\psi}(\mathbf{u})} \, |\mathbf{u}|^{-\alpha d} \, d\mathbf{u}$$

with

$$c_{\alpha,d} := \frac{\Gamma(\alpha d/2)}{\Gamma((1-\alpha)d/2) \, \pi^{(\alpha-1/2)d}} \cdot$$

Let $H \in (1/2, 1)$ and denote $\alpha := 2H - 1 \in (0,1)$. A centered Gaussian process $\mathbb{W}_H(\varphi)$, $\varphi \in \mathbb{M}_\alpha$, with covariance

$$\mathbb{E} \, \mathbb{W}_H(\varphi)\mathbb{W}_H(\psi) := c_{\alpha,d}^{-1} \, [\varphi, \psi]_\alpha = c_{\alpha,d}^{-1} \int_{\mathbb{R}^d} \int_{\mathbb{R}^d} \frac{\varphi(\mathbf{t}_1)\psi(\mathbf{t}_2)}{|\mathbf{t}_1 - \mathbf{t}_2|^{(2-2H)d}} \, d\mathbf{t}_1 d\mathbf{t}_2$$

$$= \int_{\mathbb{R}^d} \widehat{\varphi}(\mathbf{u}) \, \overline{\widehat{\psi}(\mathbf{u})} \, |\mathbf{u}|^{-\alpha d} \, d\mathbf{u}$$

is called *fractional noise integral*.

Notice that for $H = 1/2$ the *formal* plugging of $\alpha = 2H - 1 = 0$ into the latter representation of covariance yields, by Parseval identity,

$$\mathbb{E}\,\mathbb{W}_{1/2}(\varphi)\mathbb{W}_{1/2}(\psi) = \int_{\mathbb{R}^d} \widehat{\varphi}(\mathbf{u})\overline{\widehat{\psi}(\mathbf{u})}\,d\mathbf{u} = \int_{\mathbb{R}^d} \varphi(\mathbf{t})\psi(\mathbf{t})\,d\mathbf{t}.$$

Therefore, we arrive at the white noise covariance (15.5).

Remark 15.2. Notice that W_H is a self-similar process: for any $c > 0$ the process $X(\varphi) := c^{-Hd}\mathbb{W}_H\left(\varphi\left(\frac{\cdot}{c}\right)\right)$ is again a fractional noise integral.

We will explain now why $\mathbb{W}_H$ can be considered as a spacial analogue of the fractional Brownian motion. In the univariate case $d = 1, 1/2 < H < 1$, we obtain the following variance for $B_H(t) := \mathbb{W}_H(\mathbf{1}_{\{[0,t]\}})$

$$c_{\alpha,1}\,\mathbb{E}\,|B_H(t)|^2 = \int_0^t \int_0^t |t_2 - t_1|^{-(2-2H)}dt_1 dt_2,$$

$$= 2\int_0^t \int_0^{t_1} (t_1 - t_2)^{-(2-2H)}dt_2 dt_1$$

$$= 2\int_0^t \int_0^{t_1} v^{-(2-2H)}dv dt_1$$

$$= 2\int_0^t \frac{t_1^{-(1-2H)}}{2H-1}\,dt_1 = \frac{t^{2H}}{H(2H-1)}\,.$$

If $s \leq t$, we obtain

$$\mathbb{E}\,|B_H(t) - B_H(s)|^2 = \mathbb{E}\,(\mathbb{W}_H(\mathbf{1}_{\{[0,t]\}}) - \mathbb{W}_H(\mathbf{1}_{\{[0,s]\}}))^2$$

$$= \mathbb{E}\,\mathbb{W}_H(\mathbf{1}_{\{(s,t]\}})^2 = \mathbb{E}\,\mathbb{W}_H(\mathbf{1}_{\{(0,t-s]\}})^2$$

$$= \frac{(t-s)^{2H}}{H(2H-1)c_{\alpha,1}}\,,$$

and this brings us (up to a constant) back to the alternative definition of fractional Brownian motion (6.15).

We refer to [20] for further information on spacial fractional processes.

15.3   *Limit theorems for spacial load*

We will study the macroscopic behavior of W^*, i.e. the behavior of load on large parts of $\mathbb{R}^d$. To this aim, we introduce a *zoom-out* scaling $\mathbf{s} \mapsto \mathbf{s}/a$, with scaling parameter $a \to \infty$, and introduce the normalized and rescaled integral load process

$$Z_a(\varphi) := \frac{W^*\left(\varphi\left(\frac{\cdot}{a}\right)\right) - \mathbb{E}\,W^*\left(\varphi\left(\frac{\cdot}{a}\right)\right)}{b}$$

$$= \frac{1}{b}\int_{\mathcal{R}} r \int \varphi\left(\frac{\mathbf{t}}{a}\right)\mathbf{1}_{\{\mathbf{t}\in\mathbf{s}+uG\}}d\mathbf{t}\,d\widetilde{N}. \qquad (15.7)$$

Here the strength scaling factor $b = b(a, \lambda)$ depends on the arrival intensity λ and on the space scaling factor a. Precise form of $b(\cdot, \cdot)$ depends on the assumptions that we impose on the distributions F_U and F_R, – exactly as in the one-parametric setting (13.1).

15.3.1 Weak dependence: convergence to white noise integral

The following theorem extends the univariate Theorem 13.1.

Theorem 15.3. *Let* $\mathbb{E}R^2 < \infty$, $\mathbb{E}\left(U^{2d}\right) < \infty$, *and assume that* $a \to \infty$ *and* $a^d\lambda \to \infty$. *Define the scaling factor by*

$$b(a, \lambda) := \left(\lambda \, \mathbb{E}R^2 \, \mathbb{E}\left(U^{2d}\right)a^d\right)^{1/2}. \tag{15.8}$$

Let $\mathbb{W}$ *be a white noise integral. Then we have*

$$Z_a \xrightarrow{f.d.d.} \mathbb{W}$$

on $\mathbb{L}_2(\mathbb{R}^d) \cap \mathbb{L}_1(\mathbb{R}^d)$.

Proof. By using (15.4) and making the variable changes $\mathbf{t} = a\tilde{\mathbf{t}}$, $\mathbf{s} = a\tilde{\mathbf{s}}$, we may write the variance as

$$\begin{aligned}
VarZ_a(\varphi) &= b^{-2}VarW^*\left(\varphi\left(\frac{\cdot}{a}\right)\right) \\
&= \frac{\lambda \, \mathbb{E}R^2}{b^2} \int\int\left(\int_{\mathbf{s}+uG}\varphi(\mathbf{t}/a)d\mathbf{t}\right)^2 ds F_U(du) \\
&= \frac{\lambda \, \mathbb{E}R^2 \, a^{3d}}{b^2} \int\int\left(\int_{\tilde{\mathbf{s}}+uG/a}\varphi(\tilde{\mathbf{t}})\,d\tilde{\mathbf{t}}\right)^2 d\tilde{\mathbf{s}} F_U(du).
\end{aligned}$$

Notice that for fixed u and $a \to \infty$ we have

$$\int\left(\int_{\tilde{\mathbf{s}}+uG/a}\varphi(\tilde{\mathbf{t}})\,d\tilde{\mathbf{t}}\right)^2 d\tilde{\mathbf{s}} \sim (u/a)^{2d}\int\varphi(\tilde{\mathbf{s}})^2 d\tilde{\mathbf{s}}.$$

Since

$$\mathbb{E}\left(U^{2d}\right) = \int u^{2d}F_U(du) < \infty,$$

we obtain

$$VarZ_a(\varphi) \sim \frac{\lambda \, \mathbb{E}R^2 \, \mathbb{E}\left(U^{2d}\right)a^d}{b^2}\int\varphi(\tilde{\mathbf{s}})^2 d\tilde{\mathbf{s}}.$$

In particular, by choosing $b(a, \lambda)$ as given in (15.8), we obtain

$$\lim_{a\to\infty}VarZ_a(\varphi) = \int\varphi(\tilde{\mathbf{s}})^2 d\tilde{\mathbf{s}}. \tag{15.9}$$

By linearity, it follows that

$$\lim_{a\to\infty} \operatorname{cov}\left(Z_a(\varphi_1), Z_a(\varphi_2)\right)$$

$$= \lim_{a\to\infty} \frac{1}{2} \left[\operatorname{Var} Z_a(\varphi_1) + \operatorname{Var} Z_a(\varphi_2) - \operatorname{Var} Z_a(\varphi_1 - \varphi_2)\right]$$

$$= \frac{1}{2} \int \left[\varphi_1(\tilde{\mathbf{s}})^2 + \varphi_2(\tilde{\mathbf{s}})^2 - (\varphi_1(\tilde{\mathbf{s}}) - \varphi_2(\tilde{\mathbf{s}}))^2\right] d\tilde{\mathbf{s}}$$

$$= \int \varphi_1(\tilde{\mathbf{s}})\varphi_2(\tilde{\mathbf{s}}))d\tilde{\mathbf{s}}.$$

Therefore, the limiting covariance is that of a white noise integral.

It remains to check that the limit is Gaussian. We first consider here one-dimensional distributions.

Let us first assume that the function φ is bounded and has a bounded support:

$$\sup_{t\in\mathbb{R}^d} |\varphi(t)| \le M_\varphi; \tag{15.10}$$

$$\varphi(t) = 0 \text{ if } |t| > \rho_\varphi. \tag{15.11}$$

Given this, for any $\mathbf{s} \in \mathbb{R}^d$, $u > 0$ the kernel $\ell_{a,\varphi}(\mathbf{s}, u)$ defined by

$$\ell_{a,\varphi}(\mathbf{s}, u) := \int \varphi\left(\frac{\mathbf{t}}{a}\right) \mathbf{1}_{\{\mathbf{t}\in\mathbf{s}+uG\}} d\mathbf{t}, \tag{15.12}$$

satisfies

$$|\ell_{a,\varphi}(\mathbf{s}, u)| \le M_\varphi |uG| = M_\varphi u^d. \tag{15.13}$$

Moreover, the function $\ell_{a,\varphi}(\cdot, u)$ also has a bounded support: if we let $\rho_G := \sup_{y\in G} |y|$, then

$$\ell_{a,\varphi}(s, u) = 0, \text{ if } |s| > \rho_\varphi a + \rho_G u. \tag{15.14}$$

In order to prove the relation

$$Z_a(\varphi) \Rightarrow \mathbb{W}(\varphi), \tag{15.15}$$

which means that $P_{Z_a(\varphi)} \Rightarrow \mathcal{N}(0, ||\varphi||_2^2)$, where

$$||\varphi||_2^2 = \int_{\mathbb{R}^d} \varphi(\mathbf{s})^2 d\mathbf{s},$$

we will use Proposition 8.1. We already know from (15.9) that $\operatorname{Var} Z_a(\varphi) \to ||\varphi||_2^2$, as $a \to \infty$. Therefore, it only remains to verify the second assumption (8.3) of that proposition. Recall that by (15.7) and (15.8)

$$Z_a(\varphi) = \int_{\mathcal{R}} \frac{r\ell_{a,\varphi}(\mathbf{s}, u)}{\left(\lambda \, \mathbb{E}R^2 \, \mathbb{E}\left(U^{2d}\right)a^d\right)^{1/2}} d\tilde{N},$$

where the kernel $\ell_{a,\varphi}(\mathbf{s}, u)$ is given by (15.12) and $\widetilde{N}$ is a centered version of the Poisson measure N.

We must check that for any $\varepsilon > 0$ it is true that

$$\int_{\{r|\ell_{a,\varphi}|/(a^d\lambda)^{1/2}>\varepsilon\}} \frac{r^2\ell_{a,\varphi}(\mathbf{s}, u)^2}{a^d} F_R(dr)F_U(du)ds \to 0.$$

Let us first get rid of the zone $\{(\mathbf{s}, u, r) : u > a\}$. It is true that

$$\int_{\mathbb{R}^d} \ell_{a,\varphi}(\mathbf{s}, u)^2 d\mathbf{s} \le ||\varphi||_1^2 a^{2d}u^d, \tag{15.16}$$

since

$$\int_{\mathbb{R}^d} \ell_{a,\varphi}(\mathbf{s}, u)^2 d\mathbf{s}$$

$$\le \int \left(\int \left| \varphi\left(\frac{\mathbf{t}_1}{a}\right)\right| 1_{\{\mathbf{t}_1 \in \mathbf{s} + uG\}} d\mathbf{t}_1 \right) \left(\int \left| \varphi\left(\frac{\mathbf{t}_2}{a}\right)\right| 1_{\{\mathbf{t}_2 \in \mathbf{s} + uG\}} d\mathbf{t}_2 \right) d\mathbf{s}$$

$$= \int \int \left| \varphi\left(\frac{\mathbf{t}_1}{a}\right) \varphi\left(\frac{\mathbf{t}_2}{a}\right)\right| \left(\int 1_{\{\mathbf{t}_1 \in \mathbf{s} + uG\}} 1_{\{\mathbf{t}_2 \in \mathbf{s} + uG\}} d\mathbf{s} \right) d\mathbf{t}_1 d\mathbf{t}_2$$

$$= \int \int \left| \varphi\left(\frac{\mathbf{t}_1}{a}\right) \varphi\left(\frac{\mathbf{t}_2}{a}\right)\right| \left(\int 1_{\{\mathbf{s} \in (\mathbf{t}_1 - uG \cap \mathbf{t}_2 - uG)\}} d\mathbf{s} \right) d\mathbf{t}_1 d\mathbf{t}_2$$

$$= \int \int \left| \varphi\left(\frac{\mathbf{t}_1}{a}\right) \varphi\left(\frac{\mathbf{t}_2}{a}\right)\right| |uG| \, d\mathbf{t}_1 d\mathbf{t}_2$$

$$= u^d \left(\int \left| \varphi\left(\frac{\mathbf{t}}{a}\right)\right| d\mathbf{t} \right)^2 = u^d a^{2d} ||\varphi||_1^2.$$

From (15.16) it follows that

$$\int_{\{u>a\}} \frac{r^2\ell_{a,\varphi}(\mathbf{s}, u)^2}{a^d} F_R(dr)F_U(du)ds \le \mathbb{E}R^2 \, a^d ||\varphi||_1^2 \int_a^\infty u^d F_U(du)$$

$$\le \mathbb{E}R^2 \, ||\varphi||_1^2 \int_a^\infty u^{2d} F_U(du) \to 0.$$

Next, let us consider the zone $\{(s, u, r) : u \le a\}$, where we will use the bound

$$\int_{\mathbb{R}^d} \ell_{a,\varphi}(\mathbf{s}, u)^2 d\mathbf{s} \le C_\varphi a^d u^{2d}, \tag{15.17}$$

trivially following from (15.13) and (15.14).

Moreover, using inequality (15.13), we can cover the zone

$$\{(\mathbf{s}, u, r) : r|\ell_{a,\varphi}(\mathbf{s}, u)|/(a^d\lambda)^{1/2} > \varepsilon\}$$

we are interested in by a larger zone

$$\{(\mathbf{s}, u, r) : ru^d/(a^d\lambda)^{1/2} > \varepsilon/M_\varphi\},$$

and cover the latter by the union of two simpler zones

$$\{(\mathbf{s}, u, r) : r > (a^d \lambda \varepsilon^2 / M_\varphi^2)^{1/4}\}$$

and

$$\{(\mathbf{s}, u, r) : u^d > (a^d \lambda \varepsilon^2 / M_\varphi^2)^{1/4}\}.$$

In this way, we obtain the estimate

$$\int_{\left\{ \substack{r |\ell_{a,\varphi}| / (a^d \lambda)^{1/2} > \varepsilon \\ u \le a} \right\}} \frac{r^2 \ell_{a,\varphi}(\mathbf{s}, u)^2}{a^d} F_R(dr) F_U(du) d\mathbf{s}$$

$$\le \int_{\left\{ \substack{r > (a^d \lambda \varepsilon^2 / M_\varphi^2)^{1/4} \\ u \le a} \right\}} \frac{r^2 \ell_{a,\varphi}(\mathbf{s}, u)^2}{a^d} F_R(dr) F_U(du) d\mathbf{s}$$

$$+ \int_{\left\{ \substack{u > (a^d \lambda \varepsilon^2 / M_\varphi^2)^{1/4} \\ u \le a} \right\}} \frac{r^2 \ell_{a,\varphi}(\mathbf{s}, u)^2}{a^d} F_R(dr) F_U(du) d\mathbf{s}.$$

For the first integral by using (15.17) we obtain the bound

$$\int_{(a^d \lambda \varepsilon^2 / M_\varphi^2)^{1/4}}^{\infty} r^2 F_R(dr) \cdot a^{-d} \int_0^a \left(\int_{-\infty}^{\infty} \ell_{a,\varphi}(\mathbf{s}, u)^2 d\mathbf{s} \right) F_U(du)$$

$$\le \int_{(a^d \lambda \varepsilon^2 / M_\varphi^2)^{1/4}}^{\infty} r^2 F_R(dr) \cdot a^{-d} \cdot C_\varphi a^d \int_0^{\infty} u^{2d} F_U(du) \to 0,$$

due to the first factor. We used here theorem's assumptions: $\mathbb{E} R^2 < \infty$, $\mathbb{E}(U^{2d}) < \infty$, and $a^d \lambda \to \infty$.

For the second integral we obtain, still by using (13.8), the bound

$$\int_{\mathbb{R}} r^2 F_R(dr) \cdot a^{-d} \int_{(a^d \lambda \varepsilon^2 / M_\varphi^2)^{1/4}}^{a} \left(\int_{-\infty}^{\infty} \ell_{a,\varphi}(\mathbf{s}, u)^2 d\mathbf{s} \right) F_U(du)$$

$$\le \int_{\mathbb{R}} r^2 F_R(dr) \cdot a^{-d} \cdot C_\varphi a^d \int_{(a \lambda \varepsilon^2 / M_\varphi^2)^{1/4}}^{a} u^{2d} F_U(du)$$

$$\le C_\varphi \int_{\mathbb{R}} r^2 F_R(dr) \cdot \int_{(a^d \lambda \varepsilon^2 / M_\varphi^2)^{1/4}}^{\infty} u^{2d} F_U(du) \to 0,$$

due to the last factor. Again, we used here the theorem's assumptions $\mathbb{E} R^2 < \infty$, $\mathbb{E}(U^{2d}) < \infty$, and $a^d \lambda \to \infty$.

We have now proved weak convergence (15.15) for any function φ satisfying additional assumptions (15.10) and (15.11).

Since one may approximate in $\mathbb{L}_2(\mathbb{R}^d)$ any function from $\mathbb{L}_2(\mathbb{R}^d)$ by functions satisfying (15.10) and (15.11), the relation (15.15) easily extends to arbitrary function $\varphi \in \mathbb{L}_2(\mathbb{R}^d) \cap \mathbb{L}_1(\mathbb{R}^d)$ due to the asymptotic control of variances (15.9) that we have already.

Finally, due to linearity of Z_a in functional argument, the convergence of all finite-dimensional distributions of Z_a to those of white noise immediately follows via Cramér – Wold criterion. $\square$

15.3.2 *Long range dependence: convergence to a fractional noise integral*

As in the univariate case, long range dependence appears when we have sufficiently many wide grains. Here "wide" means "having width order u comparable to or larger than the observation horizon a".

Therefore, we abandon the finite variance assumption $\mathbb{E}\,(U^{2d}) < \infty$ of previous subsection and replace it with a regular variation condition

$$\mathbb{P}(U > u) \sim \frac{c_U}{u^{\gamma d}} \,, \qquad u \to \infty, \qquad 1 < \gamma < 2, \; c_U > 0, \qquad (15.18)$$

which is a spacial version of (12.6).

Notice that border condition $\gamma > 1$ provides the finite expectation $\mathbb{E}\,W^\circ(\mathbf{t}) < \infty$ while $\gamma < 2$ corresponds to the infinite variance of grain volume, i.e. $\mathbb{E}\,(U^{2d}) = \infty$.

Considering the expectation of number of grains covering simultaneously two fixed a-distant points, and requiring that this expectation must go to infinity, we arrive at the spacial version of *high intensity condition* (cf. (13.19))

$$\frac{\lambda}{a^{(\gamma-1)d}} \to \infty. \qquad (15.19)$$

We will search a limit theorem for Z_a appropriate to the long range dependence framework and choose the scaling factor b as a proper normalizer for the variance of Z_a. Recall that according to (15.7) and (15.12) we have

$$\mathbb{V}ar\,Z_a(\varphi) = \frac{\lambda}{b^2} \int_{\mathcal{R}} r^2 \ell_{a,\varphi}(s, u)^2 ds F_U(du) F_R(dr)$$

$$= \frac{\lambda\,\mathbb{E}R^2}{b^2} \int\!\!\int \ell_{a,\varphi}(s, u)^2 ds F_U(du).$$

Furthermore,

$$\ell_{a,\varphi}(s, u)^2 = \int\!\!\int \varphi\left(\frac{\mathbf{t}_1}{a}\right) \mathbf{1}_{\{\mathbf{t}_1 \in s+uG\}} \varphi\left(\frac{\mathbf{t}_2}{a}\right) \mathbf{1}_{\{\mathbf{t}_2 \in s+uG\}} \, d\mathbf{t}_1 \, d\mathbf{t}_2$$

$$= a^{2d} \int\!\!\int \varphi(\mathbf{t}_1) \mathbf{1}_{\{a\mathbf{t}_1 \in s+uG\}} \varphi(\mathbf{t}_2) \mathbf{1}_{\{a\mathbf{t}_2 \in s+uG\}} \, d\mathbf{t}_1 \, d\mathbf{t}_2$$

$$= a^{2d} \int\!\!\int \varphi(\mathbf{t}_1)\varphi(\mathbf{t}_2) \mathbf{1}_{\{s \in (a\mathbf{t}_1 - uG) \cap (a\mathbf{t}_2 - uG)\}} d\mathbf{t}_1 \, d\mathbf{t}_2.$$

By integrating this expression over the variable **s** and interchanging the integrals we come to the expression

$$a^{2d} \int \int \varphi(\mathbf{t}_1)\varphi(\mathbf{t}_2) \,|a\mathbf{t}_1 - uG) \cap (a\mathbf{t}_2 - uG)| \, d\mathbf{t}_1 \, d\mathbf{t}_2$$

$$= a^{2d} \int \int \varphi(\mathbf{t}_1)\varphi(\mathbf{t}_2) \,|(uG + a(\mathbf{t}_2 - \mathbf{t}_1)) \cap uG| \, d\mathbf{t}_1 \, d\mathbf{t}_2.$$

Performing the integration over the variable u and using regular variation assumption (15.18) leads to

$$\int_0^\infty |(uG + at) \cap uG| \, F_U(du) \sim c_U \, \gamma d \, a^{(1-\gamma)d} \int_0^\infty |(vG + t) \cap vG| \, \frac{dv}{v^{\gamma d+1}} \,,$$

as $a \to \infty$. We conclude that

$$\mathbb{V}ar Z_a(\varphi)$$

$$\sim \frac{\lambda \mathbb{E}R^2 \, c_U \gamma d a^{(3-\gamma)d}}{b^2} \int \int \varphi(\mathbf{t}_1)\varphi(\mathbf{t}_2) K_{G,\gamma}(\mathbf{t}_2 - \mathbf{t}_1) \, d\mathbf{t}_1 \, d\mathbf{t}_2, \quad (15.20)$$

where

$$K_{G,\gamma}(\mathbf{t}) := \int_0^\infty |(vG + \mathbf{t}) \cap vG| \, \frac{dv}{v^{\gamma d+1}} \,. \qquad (15.21)$$

Therefore, we must let

$$b := \left(\lambda \mathbb{E}R^2 \, c_U \, \gamma d\right)^{1/2} a^{(3-\gamma)d/2}. \qquad (15.22)$$

Such scaling yields for every φ

$$\mathbb{V}ar Z_a(\varphi) \to \int \int \varphi(\mathbf{t}_1)\varphi(\mathbf{t}_2) K_{G,\gamma}(\mathbf{t}_2 - \mathbf{t}_1) d\mathbf{t}_1 \, d\mathbf{t}_2, \qquad \text{as } a \to \infty.$$

By linearity of Z_a in functional argument it follows that

$$\text{cov}(Z_a(\varphi), Z_a(\psi))$$

$$= \frac{1}{2} \left(\mathbb{V}ar Z_a(\varphi) + \mathbb{V}ar Z_a(\psi) - \mathbb{V}ar(Z_a(\varphi) - Z_a(\psi))\right)$$

$$= \frac{1}{2} \left(\mathbb{V}ar Z_a(\varphi) + \mathbb{V}ar Z_a(\psi) - \mathbb{V}ar Z_a(\varphi - \psi)\right)$$

$$\to \frac{1}{2} \int \int [\varphi(\mathbf{t}_1)\varphi(\mathbf{t}_2) + \psi(\mathbf{t}_1)\psi(\mathbf{t}_2) - (\varphi - \psi)(\mathbf{t}_1)(\varphi - \psi)(\mathbf{t}_2)]$$

$$K_{G,\gamma}(\mathbf{t}_2 - \mathbf{t}_1) d\mathbf{t}_1 d\mathbf{t}_2$$

$$= \int \int \varphi(\mathbf{t}_1)\psi(\mathbf{t}_2) K_{G,\gamma}(\mathbf{t}_2 - \mathbf{t}_1) \, d\mathbf{t}_1 \, d\mathbf{t}_2, \qquad \text{as } a \to \infty.$$

The following theorem extends the univariate Theorem 13.4.

Theorem 15.4. *Assume that* $\mathbb{E}R^2 < \infty$ *and that for some* $\gamma \in (1, 2)$ *regular variation condition* (15.18) *and high intensity condition* (15.19) *are satisfied.*

Define the scaling factor by (15.22). *Then, as* $a \to \infty$, *we have*

$$Z_a \overset{f.d.d.}{\longrightarrow} B_{G,\gamma} \qquad \text{on } \mathbb{M}_{2-\gamma}$$

where $B_{G,\gamma}$ *is a centered Gaussian random process with covariance*

$$\mathbb{E}\, B_{G,\gamma}(\varphi) B_{G,\gamma}(\psi) = \int_{\mathbb{R}^d} \int_{\mathbb{R}^d} \varphi(\mathbf{t}_1)\psi(\mathbf{t}_2) K_{G,\gamma}(\mathbf{t}_2 - \mathbf{t}_1)\, d\mathbf{t}_1 d\mathbf{t}_2\,,$$

and the kernel $K_{G,\gamma}(\cdot)$ *is given by* (15.21).

In particular, when the grain G *is a ball, then* $B_{G,\gamma}$ *coincides with a fractional noise integral* $\mathbb{W}_H$, $H = \frac{3-\gamma}{2} \in (1/2, 1)$, *up to a numerical constant.*

Remark 15.5. One comes to the fractional noise integral not only for the balls but also for rather arbitrary spherically symmetric grains satisfying minor regularity assumptions, see [48].

Proof. First of all, let us check that the process $B_{G,\gamma}$ is well defined on $\mathbb{M}_{2-\gamma}$. Consider expression (15.21) attentively. Notice that the set $(vG + \mathbf{t}) \cap vG$ is empty whenever $v < \frac{|t|}{2\rho_G}$ where we still use the notation $\rho_G = \sup_{y \in G} |y|$. Furthermore, the trivial bound

$$|(vG + \mathbf{t}) \cap vG| \le |vG| = v^d$$

yields

$$K_{G,\gamma}(t) \le \int_{\frac{|t|}{2\rho_G}}^{\infty} v^{-(\gamma-1)d-1} dv = \frac{1}{(\gamma-1)d} \left(\frac{|t|}{2\rho_G} \right)^{-(\gamma-1)d} := C|t|^{-(\gamma-1)d}.$$

By comparing this bound with the definition of Riesz energy (15.6), where we let $\alpha := 2 - \gamma$, we see that covariance of $B_{G,\gamma}$ is finite on $\mathbb{M}_{2-\gamma}$.

Next, let us start proving convergence of one-dimensional distributions. In order to prove that for each $\varphi \in \mathbb{M}_{2-\gamma}$

$$Z_a(\varphi) \Rightarrow B_{G,\gamma}(\varphi),$$

we will use Proposition 8.1 dealing with convergence of Poisson integrals to a normal limit. We already know from (15.20) that $\mathbb{V}ar Z_a(\varphi) \to \mathbb{V}ar B_{G,\gamma}(\varphi)$, as $a \to \infty$. Therefore, it only remains to verify the second assumption (8.3) of that proposition.

Recall that by (15.7) and (15.22)

$$Z_a(\varphi) = \int_{\mathcal{R}} \frac{r\ell_{a,\varphi}(\mathbf{s}, u)}{(K\lambda)^{1/2} a^{(3-\gamma)d/2}} \, d\widetilde{N},$$

where the kernel $\ell_{a,\varphi}(\cdot, \cdot)$ is defined in (15.12), $\widetilde{N}$ is a centered version of Poisson measure N and $K := \mathbb{E}R^2 \, c_U \, \gamma d$ is a constant which is unimportant at this point. Let

$$f_a(\mathbf{s}, u, r) := \frac{r\ell_{a,\varphi}(\mathbf{s}, u)}{a^{(3-\gamma)d/2}\lambda^{1/2}} \ .$$

Checking (8.3) means to verify that for any $\varepsilon > 0$ it is true that

$$\int_{\{|f_a|>\varepsilon\}} f_a^2 F_R(dr) F_U(du) \lambda d\mathbf{s} \to 0. \qquad (15.23)$$

Fix a large $M > 0$ and split the integral into two parts:

$$\int_{\{|f_a|>\varepsilon\}} f_a^2 d\mu = \int_{\{\substack{|f_a|>\varepsilon \\ r \le M}\}} f_a^2 d\mu + \int_{\{\substack{|f_a|>\varepsilon \\ r > M}\}} f_a^2 d\mu$$

$$\le \int_{\{\substack{|f_a|>\varepsilon \\ r \le M}\}} f_a^2 d\mu + \int_{\{r>M\}} f_a^2 d\mu.$$

We show now that the first integral simply vanishes for large a. Indeed, using first inequality

$$|\ell_{a,\varphi}(s, u)| \le \|\varphi\|_1 \, a^d,$$

then applying assumption (15.19) we have, for the triplets $(\mathbf{s}, u, r)$ satisfying $r \le M$,

$$|f_a(\mathbf{s}, u, r)| \le \frac{M\|\varphi\|_1 \, a^d}{a^{(3-\gamma)d/2}\lambda^{1/2}} = M\|\varphi\|_1 \left(\frac{a^{(\gamma-1)d}}{\lambda}\right)^{1/2} \to 0,$$

which is incompatible with condition $|f_a| > \varepsilon$.

For the the second integral, we use the former bounds for the variance of $Z_a(\varphi)$, yet with R replaced with $R\mathbf{1}_{\{R>M\}}$. Hence, instead of $\mathbb{E}R^2$ the factor $\mathbb{E}\left(R^2 \mathbf{1}_{\{R>M\}}\right)$ appears. By choosing M to be large enough, we may render this expression arbitrarily small, thus (15.23) follows and convergence of one-dimensional distributions of $Z_a(\cdot)$ is verified.

Furthermore, due to linearity of Z_a in functional argument, the convergence of all finite-dimensional distributions of $Z_a(\cdot)$ to those of $B_{G,\gamma}(\cdot)$ immediately follows via Cramér – Wold criterion.

Finally, if the grain G is a ball, then the kernel $K_{G,\gamma}(\cdot)$ is spherically invariant; easy scaling argument yields

$$K_{G,\gamma}(t) = C_{G,\gamma}|t|^{-(\gamma-1)d} = C_{G,\gamma}|t|^{-(2H-2)d},$$

with appropriate constant $C_{G,\gamma}$, whenever $H = \frac{3-\gamma}{2}$. This expression coincides with the kernel of the corresponding fractional integral (Riesz energy of order $\alpha = 2 - \gamma$) up to the constant $C_{G,\gamma}$, and we are done. $\qquad\square$

15.3.3 *Concluding remarks*

Remark 15.6. Of course, spacial limit theorems extending univariate results with Lévy stable and Telecom processes as limits also exist. See [10, 14, 48], etc.

Remark 15.7. Along with the zoom-out scaling that we considered here, it is equally reasonable to consider a *zoom-in* scaling $\mathbf{s} \mapsto \mathbf{s}/a$, with $a \to 0$, in order to investigate the local (microscopic) structure of the load process. This is reasonable to do if we allow infinitely many small grains to appear in a bounded space domain. This situation may be incorporated in the model by letting the measure $F_U(du)$ to accumulate infinite mass at zero and impose regularity conditions on its lower tails.

Remark 15.8. See e.g. [114] and the references therein for further applications of Poisson random measures in telecommunication models.

Remark 15.9. We considered here the H-self-similar processes from the range $H \in [1/2, 1)$. One can obtain limit self-similar processes with wider range of parameter H, provided that the space of test functions φ is reduced appropriately, cf [10].

Notations

$a_X(\cdot)$ – expectation function of a random process X

$\mathcal{A}_X$ – sigma-field associated to a random variable or random vector X

$\mathcal{B}$ – Borel sigma-field on the real line

$\mathcal{B}^n$ – Borel sigma-field on the n-dimensional Euclidean space

$\mathcal{B}_{\mathcal{X}}$ – Borel sigma-field on a metric space $\mathcal{X}$

B^H – fractional Brownian motion

$\mathbb{C}$ – complex numbers

$\mathbb{C}[0,1]$ – space of continuous real-valued functions on the interval $[0,1]$

$\mathcal{C}(a,\sigma)$ – Cauchy distribution

CLT – central limit theorem

$\operatorname{cov}(X,Y)$ – covariance of random variables X, Y

$\mathbb{D}[0,1]$ – Skorokhod space of cadlag functions on the interval $[0,1]$

$\mathbb{E}\,X$ – expectation of a random variable X

$E_\alpha(\varphi)$ – Riesz energy

$\mathcal{E}(\cdot)$ – empirical process

fBm – fractional Brownian motion

f.d.d. – finite-dimensional distributions

F_X – distribution function of a random variable X

$f_X(\cdot)$ – characteristic function of a random variable X

$\mathcal{G}(\alpha)$ – Gamma distribution

i.i.d. – independent identically distributed (random variables, vectors, etc)

$K_X(\cdot,\cdot)$ – covariance function of a random process X

LLN – law of large numbers

$\mathbb{M}_\alpha$ – space of functions of finite Riesz energy

$N(\cdot)$ – Poisson random measure

$\widetilde{N}(\cdot)$ – centered Poisson random measure

$\mathbb{N}$ – positive integers $\{1, 2, \dots\}$

$\mathcal{N}(a, \sigma^2)$ – normal distribution

$\mathcal{N}(a, K)$ – multivariate Gaussian distribution

P_X – distribution of a random variable or random vector X

$P_{X_1, \dots, X_n}$ – joint distribution of random variables $X_1, \dots, X_n$

$P^X_{t_1, \dots, t_n}$ – finite-dimensional distribution of a process X

$p_X(\cdot)$ – distribution density of a random variable X

$\mathbb{P}$ – probability measure

$\mathcal{P}(a)$ – Poisson distribution of intensity a

$\mathcal{P}_{1,0}$ – class of distributions of Poisson integrals

$\mathcal{P}_{2,0}$ – class of infinitely divisible distributions without Gaussian component

$\mathcal{P}_{2,1}$ – class of distributions of centered Poisson integrals

$\mathbb{R}$ – set of all real numbers

$\mathbb{R}^n$ – n-dimensional Euclidean space

$\mathbb{R}_+$ – set of all non-negative real numbers

R_α – Riemann–Liouville operator

$R^\alpha(\cdot)$ – Riemann–Liouville process

$\mathbb{S}^1$ – unit circle on the plane

$\widetilde{\mathcal{S}}(c_-, c_+, \alpha)$ – strictly α-stable distribution ($\alpha \neq 1$)

$\widetilde{\mathcal{S}}(a, c, 1)$ – strictly α-stable distribution ($\alpha = 1$)

$U(\cdot)$ – Ornstein–Uhlenbeck process

$U^H(\cdot)$ – fractional Ornstein–Uhlenbeck process

$\mathbb{V}arX$ – variance of a random variable X

$W(\cdot)$ – Wiener process

$W^C(\cdot)$ – Brownian sheet (Wiener-Chentsov field)

$\mathbb{W}(\cdot)$ – white noise integral

$W^H(\cdot)$ – fractional Brownian sheet

$\mathbb{W}_H(\cdot)$ – fractional noise integral

$W^L(\cdot)$ – Lévy's Brownian motion (function)

$\overset{o}{W}$ – Brownian bridge

$W^\circ(\cdot)$ – instant workload process in a teletraffic model

$W^*(\cdot)$ – integral workload process in a teletraffic model

$\mathcal{W}(\cdot)$ – Gaussian white noise

$Y_{\gamma, R}$ – Poisson Telecom process

$Z_a(\cdot)$ – normalized workload process in a teletraffic model

$Z_a^{(m)}(\cdot)$ – normalized workload process in a micropulse model

$\mathbb{Z}$ – integer numbers $\{0, \pm 1, \pm 2, \dots\}$

$\mathcal{Z}_{\gamma,\delta}$ – stable Telecom process

$\mathcal{Z}_{\gamma,\delta}^{(m)}$ – bifractional stable process

δ_u – unit measure concentrated at a point u

$\Gamma(\cdot)$ – Gamma function

$\lambda^d(\cdot)$ – Lebesgue measure in d-dimensional Euclidean space

$\rho(X, Y)$ – correlation coefficient of random variables X, Y

$\rho_{\mathbb{C}}(\cdot, \cdot)$ - uniform distance

$\rho_{\mathbb{D},J}(\cdot, \cdot)$ - Skorokhod J-distance

$\rho_{\mathbb{D},M}(\cdot, \cdot)$ - Skorokhod M-distance

Ω – sample space

$\xrightarrow{\text{a.s.}}$ – almost sure convergence

$\xrightarrow{\text{f.d.d.}}$ – convergence in finite-dimensional distributions

$\xrightarrow{\mathbb{L}_p}$ – convergence in the mean of order p

$\xrightarrow{\mathbb{P}}$ – convergence in probability

$\Rightarrow$ – convergence in distribution (weak convergence)

$||f||_p := \left(\int |f|^p \right)^{1/p}$ – $\mathbb{L}_p$-norm of a function f

$[x]$ – integer part of x

Bibliography

[1] Adler, R. J. (1990). *An Introduction to Continuity, Extrema and Related Topics for General Gaussian Processes, Lect. Notes Inst. Math. Stat.*, Vol. 12 (IMS, Hayword).

[2] Aksenova, K. A. (2011). On stochastic teletraffic models with heavy-tailed distributions, *J. Math. Sci. (N.Y.)* **176**, 2, pp. 103-111.

[3] Araman, V. F. and Glynn, P. W. (2012). Fractional Brownian motion with $H < 1/2$ as a limit of scheduled traffic, *J. Appl. Probab.* **49**, pp. 710–718.

[4] Avram, F. and Taqqu, M. S. (1989). Probability bounds for M-Skorohod oscillations, *Stoch. Proc. Appl.* **33**, pp. 63–72.

[5] Avram, F. and Taqqu, M. S. (1992). Weak convergence of sums of moving averages in the α-stable domain of attraction, *Ann. Probab.* **20**, pp. 483–503.

[6] Bardina, X. and Es-Sebaiy, K. (2011). An extension of bifractional Brownian motion, *Commun. Stochast. Analysis* **5**, pp. 333–340.

[7] Bertoin, J. (1998). *Lévy Processes, Ser.: Cambridge Tracts in Mathematics*, Vol. 121 (Cambridge University Press).

[8] Biermé, H. and Estrade, A. (2006). Poisson random balls: self-similarity and X-ray images, *Adv. Appl. Probab.* **38**, 4, pp. 853–872.

[9] Biermé, H. and Estrade, A. (2012). Covering the whole space with Poisson random balls, *ALEA: Lat. Amer. J. Probab. Math. Statist.* **9**, 1, pp. 213–229.

[10] Biermé, H., Estrade, A., Kaj, I. (2010). Self-similar random fields and rescaled random balls models, *J. Theor. Probab.* **23**, 4, pp. 1110–1141.

[11] Billingsley, P. (1968). *Convergence of Probability Measures* (Wiley, New York).

[12] Bogachev, V. I. (1998) *Gaussian Measures, Ser. Math. Surveys and Monographs*, Vol. 62 (AMS, Providence).

[13] Breton, J.-C. and Dombry, C. (2009). Rescaled random ball models and stable self-similar random fields, *Stoch. Proc. Appl.* **119**, pp. 3633–3652.

[14] Breton, J.-C. and Dombry, C. (2011). Functional macroscopic behavior of weighted random balls model, *ALEA: Lat. Amer. J. Probab. Math. Statist.*

8, pp. 177–196.

[15] Bulinski, A. V. and Shashkin, A. P. (2007). *Limit Theorems for Associated Random Fields and Related Systems, Advanced Series on Statistical Science & Applied Probability*, Vol.10 (World Scientific, Hackensack).

[16] Chentsov, N. N. (1961). Lévy Brownian motion for several parameters and generalized white noise, *Theor. Probab. Appl.* **2**, pp. 265–266.

[17] Chentsov, N. N. and Morozova, E. A. (1967). Lévy's random fields. *Theor. Probab. Appl.* **12**, pp. 153-156.

[18] Cioczek-Georges, R. and Mandelbrot, B. B. (1995). A class of micropulses and antipersistent fractional Brownian motion, *Stoch. Proc. Appl.* **60**, pp. 1–18.

[19] Cioczek-Georges, R. and Mandelbrot, B. B. (1996). Alternative micropulses and fractional Brownian motion, *Stoch. Proc. Appl.* **64**, pp. 143-152.

[20] Cohen, S. and Istas, J. (2013). *Fractional Fields and Applications* (Springer, to appear).

[21] Cohen, S. and Lifshits, M. (2012). Stationary Gaussian random fields on hyperbolic spaces and on Euclidean spheres, *ESAIM: Probability and Statistics*, **16**, pp. 165–221.

[22] Cohen, S. and Taqqu, M. (2004). Small and large scale behavior of the Poissonized Telecom Process, *Methodol. Comput. Appl. Probab.* **6**, pp. 363-379.

[23] Csörgő, M. and Révész, P. (1981). *Strong Approximations in Probability and Statistics* (Academic Press, New York).

[24] D'Auria, B. and Resnick, S. I. (2006). Data network models of burstiness, *Adv. Appl. Probab.* **38**, pp. 373-404.

[25] D'Auria, B. and Resnick, S. I. (2008). The influence of dependence on data network models, *Adv. Appl. Probab.* **40**, 1, pp. 60-94.

[26] Dobrushin, R. L. and Major, P. (1979). Non-central limit theorem for nonlinear functionals of Gaussian fields, *Z. Wahrsch. verw. Gebiete* **50**, pp. 1-28.

[27] Dombry, C. and Kaj, I. (2011). The on-off network traffic model under intermediate scaling, *Queueing Systems* **69**, pp. 29–44.

[28] Donsker, M. (1951). An invariance principle for certain probability limit theorems, *Mem. Amer. Math. Soc.* **6**, pp. 1–12.

[29] Doukhan, P., Oppenheim, G., and Taqqu, M. S. (eds.) (2003). *Theory and Applications of Long-range Dependence* (Birkhäuser, Boston).

[30] Einmahl, U. (1989). Extensions of results of Komlós, Major, and Tusnády to the multivariate case. *J. Multivar. Anal.* **28**, pp. 20–68.

[31] Embrechts, P. and Maejima, M. (2002). *Selfsimilar Processes.* (Princeton Univ. Press, Princeton).

[32] Estrade, A. and Istas, J. (2010). Ball throwing on spheres, *Bernoulli* **16**, pp. 953–970.

[33] Faraut, J. and Harzallah, K. (1974). Distances hilbertiennes invariantes sur

un espace homogène, *Ann. Inst. Fourier (Grenoble)* **24**, pp. 171-217.

[34] Fasen, V. (2010). Modeling network traffic by a cluster Poisson input process with heavy and light-tailed file sizes, *Queueing Systems* **66**, pp. 313–350.

[35] Gaigalas, R. (2006). A Poisson bridge between fractional Brownian motion and stable Lévy motion, *Stoch. Proc. Appl.* **116**, pp. 447-462.

[36] Gaigalas, R. and Kaj, I. (2003). Convergence of scaled renewal processes and a packet arrival model, *Bernoulli* **9**, pp. 671–703.

[37] Gangolli, R. (1967). Positive definite kernels on homogeneous spaces and certain stochastic processes related to Lévy's Brownian motion of several parameters, *Ann. Inst. H. Poincaré, Ser. B* **3**, pp. 121-226.

[38] Gross, D. et al. (2008). *Fundamentals of Queueing Theory* (Wiley, New York).

[39] Guerin, C. A. et al. (2003). Empirical testing of the infinite source Poisson data traffic model, *Stoch. Models* **19**, pp. 151–200.

[40] Houdré, C. and Villa, J. (2003). An example of infinite dimensional quasi-helix, *Stochastic Models, Contemporary Mathematics* Vol. **336**, pp. 195–201.

[41] Ibragimov, I. A. and Linnik, Yu. V. (1971). *Independent and Stationary Sequences of Random Variables* (Wolters-Noordhof).

[42] Istas, J. (2005). Spherical and hyperbolic fractional Brownian motion, *Electron. Commun. Probab.* **10**, pp. 254-262.

[43] Istas, J. (2006). On fractional stable fields indexed by metric spaces, *Electron. Commun. Probab.* **11**, pp. 242-251.

[44] Kaj, I. (2002). *Stochastic Modeling in Broadband Communications Systems, SIAM Monographs on Mathematical Modeling and Computation* Vol.8 (SIAM, Philadelphia).

[45] Kaj, I. (2005). *Limiting fractal random processes in heavy-tailed systems, Fractals in Engineering, New Trends in Theory and Applications,* J. Levy-Vehel, E. Lutton (eds.), pp. 199–218 (Springer-Verlag, London).

[46] Kaj, I. (2006). *Aspects of Wireless Network Modeling Based on Poisson Point Processes, Fields Institute Workshop on Applied Probability* (Carleton University, Ottawa).

[47] Kaj, I. (2010). *Stochastic Modeling for Engineering Studies, Lecture Notes, Department of Mathematics,* (Uppsala University).

[48] Kaj, I., Leskelä, L., Norros, I., and Schmidt, V. (2007). Scaling limits for random fields with long-range dependence, *Ann. Probab.* **35**, pp. 528–550.

[49] Kaj, I. and Taqqu, M. S. (2008). *Convergence to fractional Brownian motion and to the Telecom process: the integral representation approach, In and Out of Equilibrium. II., ser.: Progress in Probability,* Vol. 60,(Birkhäuser, Basel), pp. 383–427.

[50] Kingman, J. F. C. (1993). *Poisson Processes, Oxford Stud. Prob.,* Vol. 3, (Oxford University Press, New York).

[51] Kolmogorov, A. N. (1940). Wienersche Spiralen und einige andere interes-

sante Kurven im Hilbertschen Raum, *C. R. (Doklady) Acad. URSS (NS)* **26**, pp. 115–118.

[52] Komlós, J., Major, P., and Tusnády, G. (1975). An approximation of partial sums of independent RV'-s and the sample DF.I, *Z. Wahrsch. verw. Gebiete* **32**, pp. 111–131.

[53] Komlós, J., Major, P., and Tusnády, G. (1976). An approximation of partial sums of independent RV'-s and the sample DF.II, *Z. Wahrsch. verw. Gebiete* **34**, pp. 34–58.

[54] Kurtz, T. G. (1996). *Limit theorems for workload input models, Stochastic Networks, Theory and Applications* (Clarendon Press, Oxford) Kelly, F. P., Zachary, S. and Ziedins, I. (eds.), pp. 119–140.

[55] Kyprianou, A. E. (2006). *Introductory Lectures on Fluctuations of Lévy Processes with Applications, Ser.: Universitext* (Springer-Verlag, Berlin).

[56] Lei, P. and Nualart, D. (2009). A decomposition of the bifractional Brownian motion and some applications, *Statist. Probab. Letters* **79**, pp. 619–624.

[57] Levy, J. B. and Taqqu, M. S. (2000). Renewal reward processes with heavy-tailed interrenewal times and heavy-tailed rewards, *Bernoulli* **6**, pp. 23–44.

[58] Lévy, P. (1965). *Processus stochastiques et mouvement brownien (2-me edition)* (Gautier-Villars, Paris).

[59] Lieb, E. H. and Loss, M. (1997). *Analysis, Ser.: Graduate Studies in Math.*, Vol. 14 (AMS, Providence).

[60] Lifshits, M. A. (1995). *Gaussian Random Functions* (Kluwer, Dordrecht).

[61] Lifshits, M. A. (2012). *Lectures on Gaussian Processes, Ser.: Springer Briefs in Mathematics* (Springer).

[62] Lifshits, M. A. and Simon, T. (2005). Small deviations for fractional stable processes, *Ann. Inst. H. Poincaré, Ser. B* **41**, pp. 725–752.

[63] Linde, W. (1983) *Infinitely Divisible and Stable Measures on Banach Spaces, Ser.:Teubner-Texte zur Mathematik*, Vol. 58 (B.G. Teubner Verlagsgesellschaft, Leipzig).

[64] Major, P. (1979). An improvement of Strassen's invariance principle, *Ann. Probab.* **7**, pp. 55–61.

[65] Major, P. (1981). *Multiple WienerIto Integrals: With Applications to Limit Theorems, Lecture Notes in Math.* **849** (Springer, Berlin).

[66] Mandelbrot, B. B. and van Ness, J. (1968). Fractional Brownian motions, fractional noises and applications, *SIAM Rev.* **10**, pp. 422–437.

[67] Mandjes, M. (2007). *Large Deviations for Gaussian Queues* (Wiley, New York).

[68] Marinucci, D. and Robinson, P. M. (1999). Alternative forms of fractional Brownian motion, *J. Stat. Plan. Infer.* **80**, pp. 111–122.

[69] Marouby, M. (2011). Micropulses and different types of Brownian motion, *J. Appl. Probab.* **48**, pp. 792–810.

[70] Maulik, K. and Resnick, S. I. (2003). The self-similar and multifractal nature

of a network traffic model, *Stoch. Models* **19**, pp. 549–577.

[71] Mikosch, T., Resnick, S. I., Rootzén, H., and Stegeman, A. W. (2002). Is network traffic approximated by stable Lévy motion or fractional Brownian motion? *Ann. Appl. Probab.* **12**, pp. 23–68.

[72] Molchan, G. M. (2003). *Historical comments related to Fractional Brownian motion, Theory and Applications of Long Range Dependence* (Birkhäuser, Boston), Doukhan, P. et al (eds.), pp. 39-42.

[73] Newman, C.M. and Wright, A.L.(1981). An invariance principle for certain dependent sequences, *Ann. Probab.* **9**, pp. 671-675.

[74] Newman, C.M. and Wright, A.L.(1982). Associated random variables and martingale inequalities, *Probab. Theor. Relat. Fields* **59**, pp. 361-371.

[75] Norros, I. (1994). A storage model with self-similar input. *Queueing Syst.* **16**, pp. 387–396.

[76] Norros, I. (1995). On the use of fractional Brownian motion in the theory of connectionless networks, *IEEE J. on Selected Areas in Comm.* **13**, pp. 953–962.

[77] Norros, I. (2005). Teletraffic as a stochastic playground, *Scand. J. Statist.* **32**, pp. 201-215.

[78] Peccati, G., Solé, J. L., Taqqu, M. S., and Utzet, F. (2010). Stein's method and normal approximation of Poisson functionals, *Ann. Probab.* **38**, 2, pp. 443–478.

[79] Peccati, G. and Taqqu, M. S. (2011). *Wiener Chaos: Moments, Cumulants and Diagrams, Bocconi & Springer Series* **1** (Springer, Milan).

[80] Pipiras V. and Taqqu, M. S. (2000). The limit of a renewal-reward process with heavy-tailed rewards is not a linear fractional stable motion, *Bernoulli* **6**, pp. 607–614.

[81] Pipiras, V. and Taqqu, M. S. (2003). *Fractional calculus and its connections to fractional Brownian motion, Theory and Applications of Long Range Dependence* (Birkhäuser, Boston), Doukhan, P. et al (eds.), pp. 165-201.

[82] Pipiras, V., Taqqu, M. S., and Levy, J. B. (2004). Slow, fast and arbitrary growth conditions for renewal reward processes when the renewals and the rewards are heavy-tailed, *Bernoulli* **10**, pp. 121–163.

[83] Pollard, D. (1984). *Convergence of Stochastic Processes* (Springer, New York).

[84] Prokhorov, Yu. V. (1956). Convergence of random processes and limit theorems in probability theory, *Theor. Probab. Appl.* **1**, pp. 157–214.

[85] Resnick, S. I. (2007). *Heavy Tail Phenomena: Probabilistic and Statistical Modeling* (Springer, New York).

[86] Rosenblatt, M. (1961). Independence and dependence, *Proc. Fourth Berkeley Symp. Math. Statist. Probab.*, pp. 411-433 (Univ. California Press).

[87] Rosenkrantz, W. A. and Horowitz, J. (2002). The infinite sourse model for internet traffic: statistical analysis and limit theorems, *Methods and Appli-*

cations of Analysis **9**, pp. 445–462.

[88] Sakhanenko, A. I. (1984). *Rate of convergence in the invariance principles for variables with exponential moments that are not identically distributed, Trudy Inst. Mat. SO AN SSSR* (Nauka, Novosibirsk) **3**, pp. 4–49 (Russian).

[89] Samorodnitsky, G. and Taqqu, M. S. (1994). *Stable non-Gaussian Random Processes: Stochastic Models with Infinite Variance* (Chapman&Hall, New York–London).

[90] Sato, K. (1999). *Lévy Processes and Infinitely Divisible Distributions* (Cambridge University Press).

[91] Seneta, E. (1976). *Regularly varying functions, Lect. Notes Math.*, Vol. 508 (Springer, Berlin).

[92] Shao, Q.-M. (1995). Strong approximation theorems for independent random variables and their applications, *J. Multivar. Anal.* **52**, pp. 107–130.

[93] Shiryaev A. N. (1996). *Probability, Ser.: Graduate Texts in Mathematics*, Vol. 95 (Springer, New York).

[94] Skorokhod, A.V. (1956). Limit theorems for stochastic processes, *Theor. Probab. Appl.* **1**, pp. 261–290.

[95] Skorokhod, A.V. (1957). Limit theorems for stochastic processes with independent increments, *Theor. Probab. Appl.* **2**, pp. 138–171.

[96] Strassen, V. (1964). An invariance principle for the law of iterated logarithm, *Z. Wahrsch. verw. Gebiete* **3**, pp. 211–226.

[97] Takenaka, S. (1991). Integral-geometric construction of self-similar stable processes, *Nagoya Math. J.* **123**, pp. 1-12.

[98] Takenaka, S., Kubo, I., and Urakawa, H. (1981). Brownian motion parametrized with metric space of constant curvature, *Nagoya Math. J.* **82**, pp. 131-140.

[99] Taqqu, M. S. (1975). Weak convergence to fractional Brownian motion and to the Rosenblatt process, *Z. Wahrsch. verw. Gebiete* **31**, pp. 287-302.

[100] Taqqu, M. S. (1979). Convergence of integrated processes of arbitrary Hermite rank, *Z. Wahrsch. verw. Gebiete* **50**, pp. 53-83.

[101] Taqqu, M. S. (2002). The modeling of Ethernet data and of signals that are heavy-tailed with infinite variance, *Scand. J. Statist.* **29**, pp. 273–295.

[102] Taqqu, M. S. (2003). *Fractional Brownian motion and long range dependence, Theory and Applications of Long Range Dependence* (Birkhäuser, Boston), Doukhan, P. et al (eds.), pp. 5-38.

[103] Taqqu, M. S. (2013). Benoît Mandelbrot and fractional Brownian motion, *Statist. Sci.* **28**, pp. 131-134.

[104] Tudor, C. A. and Xiao, Y. (2007). Sample path properties of bifractional Brownian motion, *Bernoulli* **13**, pp. 1023-1052.

[105] van der Waart, A. W. and Wellner, J. (1996). *Weak Convergence and Empirical Processes: with Applications to Statistics Springer Series in Statistics* (Springer, New York).

[106] Wentzell, A. D. (1981). *A Course in the Theory of Stochastic Processes* (McGraw–Hill, New York).

[107] Whitt, W. (2002). *Stochastic-Process Limits, Springer Series in Operation Research* (Springer, New York).

[108] Willinger, W., Paxson, V., Riedli, R. H., and Taqqu, M. S. (2003). *Long range dependence and data network traffic, Theory and Applications of Long Range Dependence* (Birkhäuser, Boston), Doukhan, P. et al (eds.), pp. 373–408.

[109] Willinger, W., Paxson, V., Taqqu, M. S. (1998). *Self-similarity and heavy tails: structural modelling of network traffic, A Practical Guide to Heavy Tails. Statistical Techniques and Applications*, Adler R. J. et al (eds.), pp. 27-53 (Birkhäuser, Boston).

[110] Willinger, W., Taqqu, M. S., Leland, M., and Wilson, D. (1995). Self-similarity in high-speed packet traffic: analysis and modelling of Ethernet traffic measurements, *Statist. Sci.* **10**, pp. 67-85.

[111] Willinger, W., Taqqu, M. S., Leland, M., and Wilson, D. (1997). *Self-similarity through high variability: statistical analysis of Ethernet LAN traffic at the source level, IEEE/ACM Trans. Networking* 5, pp. 71-96.

[112] Wolpert, R. L. and Taqqu, M. S. (2005). Fractional Ornstein-Uhlenbeck Lévy processes and the Telecom process: upstairs and downstairs, *Signal Processing* **85**, pp. 1523-1545.

[113] Yaglom, A. M. (1997). *Correlation Theory of Stationary and Related Random Functions* (Springer, Berlin).

[114] Yang, X. and Petropulu, A. P. (2003). Co-channel interference modelling in a Poisson field of interferers in wireless communications, *IEEE Trans. Signal Processing*, **51**, pp. 64–76.

[115] Zaitsev, A. Yu. (1998). Multidimensional version of the results of Komlós, Major, and Tusnády for vectors with finite exponential moments, *ESAIM: Probability and Statistics*, **2**, pp. 41–108.

[116] Zaitsev, A. Yu. (2008). Estimates of the rate of strong Gaussian approximations for sums of i.i.d. random vectors, *J. Math. Sci.*, **152**, pp. 875–884.

[117] Zolotarev, V. M. (1986). *One-dimensinal Stable Distributions, Ser.: Translations of Mathematical Monographs*, Vol. 65 (American Mathematical Society, Providence).

[118] Zukerman, M. (2012). *Introduction to Queueing Theory and Stochastic Teletraffic Models*, http://www.ee.cityu.hk/~zukerman/classnotes.pdf .

Index

Printed in the United States
By Bookmasters